# Contents

# Meet the Red Fox

**Hero or villain? Few animals divide opinion like the Red Fox. This remarkable wild canine has lived cheek by jowl with people across the northern hemisphere for thousands of years. Celebrated by some for its resourcefulness and lush red coat, reviled by others for killing livestock and raiding bins, it has worked its way deep into culture, leaving fact sometimes hard to separate from fiction. However, behind the myth, folklore and lurid headlines lies a remarkable natural history success story.**

One reason why the Red Fox looms so large in our collective consciousness is its sheer visibility. Ironically, while the urban, industrial and agricultural transformations of our modern landscape have driven many wild British species into decline, Foxes have become steadily more conspicuous and successful. Indeed, for the average British suburbanite, this unmistakable wild animal – our boldest terrestrial predator – is probably the most commonly seen native mammal. While the urban myth persists that in London you are never more than a yard from a Rat, it is a safe bet that the average Londoner lays eyes on a Fox far more frequently.

Science tells us that the Red Fox is a member of the *Vulpes* genus of 'true foxes', one of 10 genera in the dog family Canidae. It is the only species of wild canid found in the UK and enjoys the widest natural distribution of any non-human land mammal on the planet. Science also explains how the species' great versatility, including a catholic diet and broad habitat tolerance, has allowed it to survive and adapt where other carnivores have failed – not least in the hearts of our cities.

Science, of course, is not the whole story. Something about this charismatic animal – a creature that has, for better or worse, become the living embodiment of 'wily' – has driven it deep into our culture, imagination and even politics. To understand how this has happened, we must take a closer look at the animal itself.

**Opposite:** Best foot forward: a Red Fox on a mission.

**Below:** An emblem depicting a Fox and a goose in the town hall in Vohenstrauss, Bavaria, Germany.

# Worldwide wanderer

**Above:** A warm winter coat protects the Red Fox in cold climates.

The Red Fox (*Vulpes vulpes*), known in the UK simply as the Fox, is the most widespread and numerous of about 22 species around the world that bear the name fox (see also page 23). It is found not only throughout the British Isles and mainland Europe, but also ranges east through Asia as far as Japan, west across North America to Alaska, and south to India, the Middle East and the northern fringes of Africa. To the north the frozen Arctic tundra provides a natural range limit, and the species is absent from many Arctic islands, including Greenland, although climate change may already be helping it to surmount this barrier.

Add to this natural range the Fox's successful colonisation of Australia, where it was introduced during the mid-1800s, and the total area over which it roams adds up to some 70 million sq km (more than 27 million sq miles). This gives the Red Fox, after our own species, the most extensive natural range of any land mammal on the planet. In other words, 'our' Fox – an animal inseparable from the British notion of the British countryside – is a highly cosmopolitan species. Indeed, some scientists have proposed more than 48 subspecies around the globe.

**Below:** The worldwide distribution of the Red Fox.

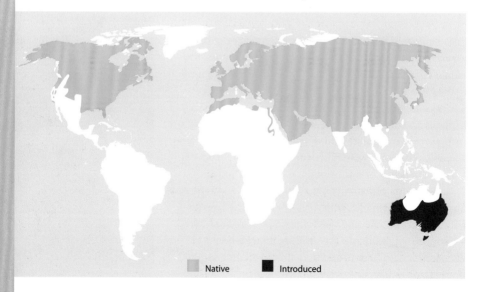

Native    Introduced

The key to this global success is the Fox's ability to get on in almost any habitat. It occurs in forests, mountains, moorland, grassland, farmland, urban areas and desert fringes, ranging from sea level up to an altitude of 3,000m (9,845ft), sometimes more. Its essential needs are food and denning sites, and Foxes tend to fare best in mixed landscapes – those comprising, for example, a hotchpotch of woodland, farmland and scrub – where they find plenty of both. Thus homogenous habitats, such as dense forests or barren uplands, may support just one Red Fox per 4–5 sq km (1½–2 sq miles), whereas more varied ones, such as farmland and deciduous woodland, typically support 1–2 Foxes per square kilometre.

The last UK Red Fox census, taken in 1999–2000, suggested a stable overall population of some 230,000 animals, before cubs are born. This averages out at roughly two Foxes per square kilometre across the country. However, population densities vary hugely by habitat. Remote Scottish hill country may support just one Fox per 30 sq km (11½ sq miles), whereas urban areas may support 4–5 Foxes in just a single square kilometre. Indeed, during the early 1990s, Bristol city centre boasted an astonishing 37 Foxes per square kilometre – the highest population density ever recorded anywhere.

**Above:** Graveyards offer plentiful refuges for urban Red Foxes.

**Below:** Red Foxes also thrive in open, arid habitats.

**Above:** Aquatic habitats offer rich pickings for the Red Fox.

**Below:** Foxes in mountainous regions tend to be larger than their lowland cousins.

# A dog with cat delusions?

Wherever you find the Red Fox, you will not struggle to identify it. The largest of the Fox species, it is roughly the size of a large, tall cat – although its thick fur and leggier stature can make it appear bigger, and as a result sizes are often exaggerated. In the UK, males have an average head and body length of 67–72cm (26–28in), with the tail one-third as long again, and tip the scales at 4–8kg (9–18lb), typically 6.5kg (14lb). Females are a little smaller, with a head and body length of 62–68cm (24–27in) and a typical weight of 5.5kg (12lb).

Foxes, in general appearance, look like smallish, slim dogs – which is, of course, exactly what they are. They have the lightweight skeleton and slim legs, with four-toed pads and non-retractable claws (five on the forefeet and four on the hind feet), that are a feature of pretty much all canines. They also have a dog's large, upright ears and long, thin, whiskered muzzle – broader in males than in females; a set of 42 teeth; and a splendid bushy tail, known as the brush, which accounts for up to 40 per cent of an individual's overall length and varies in thickness according to its state of health and the time of year.

**Below:** A sleeping Fox curls its tail around its body, much like a cat.

**Above:** A Red Fox will happily climb for a better vantage point.

**Below:** Hind legs serve for a scratch behind the ear.

Watching a Fox move you could be forgiven for thinking it had something of a cat in its genes. Rather than trotting with the bold, confident gait of most dogs, when on the hunt it moves with an almost feline caution and stealth, tripping, stalking and pouncing after its prey. Like a cat, a Fox sits and sleeps with its tail curled around its body, twitches its tail-tip to allow its cubs to practise

hunting and will even fish out of a pond with a front paw. It is also very capable of climbing trees, roofs and walls, and can scramble over a 2m (6½ft) high fence with ease. These feline qualities, however, arise from convergent evolution – the Fox has come to occupy a cat-like ecological niche, and its anatomy and behaviour have adapted accordingly.

**Above:** A Fox and a cat display similar body language – arching their backs – as they negotiate a territorial encounter.

# Boy or girl?

A male Fox is known as a dog. A female is known as a vixen. As is the case with other dogs, there is little sexual dimorphism – physical difference between the sexes – and telling them apart is not always straightforward. Many popular distinctions are erroneous: for example, a white tip to the tail is not exclusive to vixens; females do not have thinner necks or hold their heads higher than dogs, and cubs born with golden fur are not necessarily female. Dog Foxes are larger and heavier than vixens, weighing on average 17 per cent more, but this disparity is hard to make out in the field unless you see the two animals together.

There are more differences in the sexual characteristics: a male has cream-coloured fur around the scrotum; a female has visible teats, although these are only visible during the breeding season. If you have a Fox skull and a pair of calipers in your hand, you can make a fairly accurate judgement of the sex based on a few key measurements, including the length of the canines – but this method is not much use when watching one in your back garden.

Today most field naturalists agree that the best practical clue is in the shape of the face: mature dog Foxes have broader heads and thicker muzzles than vixens. Thus the face of

**Above:** Which is which? Male and female Red Foxes can be hard to distinguish in the field.

a dog Fox, seen from front on, has more of a 'W' shape, as opposed to the 'V' shape of the narrow-muzzled vixen. Even this visual clue is not foolproof, however; young males have narrow, vixen-like faces that broaden as they grow older.

**Below:** A dog Fox (left) has a noticeably broader face than a vixen (right).

# Size variations

Worldwide, Red Foxes range in head and body length from 45 to 90cm (18–35in) and in weight from 3 to 14kg (7–31lb). Typically they follow Bergmann's Rule – that is, individuals from populations in colder, more northerly latitudes are larger than those found further south. Particularly big Foxes occur in Scandinavia, Siberia and Alaska, whereas those in southern California, USA, and Mediterranean regions tend to be on the dinkier side. A popular belief that Foxes grow larger in urban areas than in the countryside due to the surfeit of food is not born out by the data.

**Below:** Foxes of southern, arid regions tend to be small and lightly built.

**Bottom:** Foxes of colder and more northerly regions are larger and thicker-furred.

# Whopping foxes

Traditionally an outsized individual Fox is a source of some pride to the fox-hunter or gamekeeper who bagged it, and records abound. As is the case with anglers' tales, however, some of the records are hard to verify and exaggeration is rife. Unconfirmed sightings by members of the public of super-sized Foxes running wild are also notoriously unreliable. Many can be put down to mistaken identity – which is hardly surprising, given that several dog breeds, including long-haired German Shepherds and red Huskies, can look very Fox-like from a distance. Whoppers do occur, however, and March 2012 proved a red-letter month in this respect, with the UK record being broken twice within a few days. The first record was of a Fox shot by a gamekeeper near East Grinstead, East Sussex, which weighed 15.8kg (35lb) and measured 130cm (4¼ft). The second was of an enormous dog Fox shot on a farm in Aberdeenshire, Scotland, which measured 145cm (4¾ft) from nose to tail-tip and weighed 17.2kg (38lb). The dimensions of the latter, which was reputedly attacking lambs, make it roughly three times bigger than an average UK Fox and closer in size to a Coyote. It is comfortably the largest Fox recorded in the UK.

**Below:** An exceptionally large male Red Fox shot in the Cairngorm Mountains, Scotland.

# A coat of many colours

If you cannot identify a Red Fox by its size, shape and the way in which it moves, then its colour should give the game away. The thick coat varies from white to black, but is predominantly a rich reddish-brown, as you would expect from the animal's name. In peak condition, during the winter months of November to February, the pelt has a flame-like lustre – especially when seen against white snow – and it is this rich hue that has made the animal so prized by furriers around the world.

The variations in a Fox's coat colour are known as morphs. They are genetically inherited and, to some extent, reflect geography. Red is the dominant morph, accounting for some 60 per cent of Foxes worldwide and much the most common across the UK. The precise hue varies from auburn to orange, with the tail generally less colourful than the body. There are usually black or dark brown markings around the eyes, muzzle and back of the ears, and black 'socks' on the paws and lower legs. A white bib extends up the neck and throat to the lower jaw and lower half of the

**Above:** Red Foxes in North America tend to have a brighter, more orange coat than those in the UK, with bolder black markings.

**Above:** Silver Foxes are most common in the far north.

**Below:** Cross Foxes, with mottled patterning, are rare in the UK.

muzzle, and the belly varies from white to slate-grey. Most Red Foxes have a prominent white tail-tip – or tag – which is visible in cubs before they are born.

Similar to red-morph Foxes are 'cross Foxes', which make up about 25 per cent of the population worldwide, but are rare in the UK. These have a black or dark brown line running along the spine and another across the shoulders and down the front legs, leaving a distinctive 'cross' mark on the shoulders. 'Silver Foxes' make up some 10 per cent of the Red Fox population worldwide, being most common in cold northern areas such as Alaska and Siberia. They have black fur frosted with white, giving them

an overall appearance that varies from black to silver-grey. Jet-black, melanistic Foxes are very rare in Europe, although they are occasionally recorded in the UK.

A Fox's coat consists of a layer of short, fine grey underfur that provides insulation by trapping air close to the skin, and a layer of longer guard hairs that carry the melanin pigmentation which gives the coat its colour. Unsurprisingly, animals that live in colder, more northerly climates tend to have the longest coats. Foxes moult progressively throughout the summer, often appearing quite tatty in late spring and early summer, and grow back their underfur during autumn.

**Above:** A typical northern European Red Fox.

**Below:** A pale morph Red Fox.

# Senses and sensitivity

Watch a Fox foraging – nose to the ground, radar ears rotating, a swift glance up at the slightest noise – and you will not be surprised to learn that this animal is equipped with a powerful battery of senses. These not only enable it to snare prey, but also alert it to danger, allow it to communicate with other foxes and, in many other ways, facilitate its journey through the world.

## Vision

**Above:** A Fox cub is born with blue eyes.

**Below:** A layer of reflective cells causes a Fox's eyes to shine when illuminated at night.

A Fox's eyes, like those of all predators, sit at the front of the skull. This allows it the binocular vision required for judging distance and seizing prey. Cubs are born with blue irises, which change to amber after 4–5 weeks, once the Fox is old enough to produce the yellow pigment lipochrome. Unlike in larger dogs, the pupils are vertical ovals, like those of a cat. This is typical of predators that hunt across many light conditions – both by day and night – allowing them more precisely to regulate the amount of light entering the eye. Also like cats and other primarily nocturnal hunters, Foxes have a layer of reflective cells behind the retina called the tapetum lucidum, which reflects light back into the eye, further improving vision in low light conditions. This is what causes a Fox's eyes to shine silver when illuminated at night in a torch beam or headlights.

In terms of overall visual acuity, Foxes have good close-range vision, enabling them to move at speed through tangled foliage and pinpoint small prey, but they are not as sharp-eyed at longer range, when their relatively low-resolution imaging and poor colour vision make it harder to identify shapes by contrast. Spotting things generally depends on movement. Thus a Fox may fail to see a small stationary object – even walking right past a crouching Rabbit.

**Above:** A Fox generally detects prey by its movement.

## Hearing

Given the limitations to its vision, a Fox needs back-up from its other senses. It has excellent hearing and is especially sensitive to low-frequency sounds, such as the rustling of prey in grass. Each ear can rotate 150 degrees in a single direction in order to help pinpoint the source of a sound – important not only for hunting but also for communication, in which sound is critical.

## Smell

Smell also plays a vital role, especially in social behaviour when it comes to sniffing out other Foxes and identifying territorial boundaries. Foxes are known to dig up carcasses from below the soil or beneath deep snow, and those who work with them often marvel at their ability to distinguish between different humans – freezing at the scent of a stranger – by smell alone. A Fox keeps its nose wet not only to help detect smells but also to help assess wind direction.

**Above:** A nose to the ground helps detect scent.

## Touch

A final string to this sensory bow is a fine sense of touch. The long vibrissae (whiskers) on a Fox's muzzle are buried three times deeper in the skin than other hairs, and attach to special nerve cells that are extremely sensitive. When a Fox is manoeuvring in dense undergrowth or poor light, these whiskers help it to build up a map of its surroundings, and when a prey item is too close for the eyes to focus, they help to direct its bite. The fur between a Fox's pads on the sole of its foot is also extremely sensitive to touch and probably helps it to navigate tricky surfaces such as tree branches and loose scree.

**Above:** Good hearing is critical to Foxes.

# Ancestors and Relatives

The Red Fox, like all foxes, is a dog. That much is straight-forward. What exactly makes a Fox, however, and how Foxes relate to other members of the dog family, is not quite so simple and requires reaching back into prehistory. To understand how the Red Fox became the animal we know today, and where Foxes take their place in the grand taxonomic filing cabinet of the animal kingdom, we must return to when dogs first appeared on the prehistoric world stage. The journey from that point to the species we know today has been long and circuitous.

# Meet the ancestors

Modern dogs – the 35 species that inhabit the earth today – make up the family Canidae. This, along with the likes of cats, weasels and bears, is one of some 16 families of mammal that form the order Carnivora. Mammals share certain key features, including body hair, three bones in the middle ear, a neocortex region of the brain and mammary glands in the females, from which they suckle their live-born young. There are a few exceptions – weirdos such as the Platypus and echidnas, for example, which lay eggs – but dogs are not among them. Man's best friend and its wilder relatives are thus both typical mammals and typical carnivores.

'Carnivore' may mean 'eating flesh', but not all members of the order Carnivora have grasped this. Some, such as certain bears and civets, eat much more in the way of fruits and/or insects than meat, while the bamboo-eating Giant Panda is entirely vegetarian. Nonetheless, all carnivores are descended from meat-eating ancestors and share a basic body plan that is adapted for hunting, killing

**Opposite:** A Red Fox among Red-crowned Cranes in Hokkaido, Japan.

**Below:** A Large-spotted Genet is similar in size and appearance to the earliest carnivores.

**Above:** *Miacis*, an Eocene mammal from Messel, Germany, was one of the earliest known carnivores.

**Above:** A Red Fox skull exhibits the typical dentition of a carnivore.

**Below:** African Wild Dogs are built for running in open grassland.

and processing flesh. Typical carnivore attributes include claws and sharp canine teeth for gripping or killing prey, carnassial teeth for shearing through flesh and forward-facing eyes for focusing on a target.

The ancestors of all carnivores are thought to have been tree-dwelling, civet-like creatures called myacids, which appeared during the mid-Eocene epoch some 42 million years ago. Their long, lithe, short-limbed body was the basic template upon which natural selection subsequently fashioned numerous carnivore variations – from the bone-crushing jaws of a hyena to the webbed feet of a Sea Otter or the sprinter's limbs of a Cheetah.

The first wave of carnivore evolution produced two basic groups: the cat-like carnivores and the dog-like carnivores. The latter are thought to have evolved in North America, with the earliest dog probably being something like the Coyote-sized *Prohesperocyon wilsoni*, which lived on the grasslands during the late Eocene, some 36 million years ago. The dog lineage split into three subfamilies, of which, by the late Miocene epoch, some 10 million years ago, just one – the Caninae – remained. By eight million years ago these animals had crossed the prehistoric land bridge of the Bering Strait into Eurasia.

The grassland habitat of the first canines was pivotal to the development of 'dogness'. Dogs, typically, are specialists in cooperative coursing, adapted for chasing prey through open terrain. Their key modifications to the carnivore template were a slim, long-limbed body built for long-distance chases, and the social skills required to assist one another in the chase. Dogs may not have the explosive spring of a cat or the brute strength of a bear, but they do not need them. Their tricks are stamina, persistence and teamwork.

# Enter the foxes

**Above:** A fox (left) has a narrower, more pointed muzzle than a wolf (right).

Around 10 million years ago, before leaving North America, the ancestral Caninae lineage had already split into two major groupings: the wolf-like dogs (*Canis*) and the fox-like dogs (*Vulpes*). The latter were smaller and more delicately built than the wolf-like dogs. These basic distinctions persist today. Typical fox features also include a bushy tail and a triangular face, with pointed ears and a sharp, slightly upturned muzzle. True foxes (*Vulpes* genus) have vertical oval pupils, like those of cats, rather than round ones like those of the wolf-like or true dogs.

Once foxes had reached the Old World, new species continued to evolve – first in Africa, then spreading to Europe via a Mediterranean land bridge around the late Miocene epoch, some seven million years ago. Today's Red Fox (*Vulpes vulpes*) and Arctic Fox (*V. lagopus*) are both thought to have descended from the Eurasian Red Fox (*V. alopecoides*), a prehistoric species that lived in southern Europe at around the end of the Pliocene epoch 2.6 million years ago.

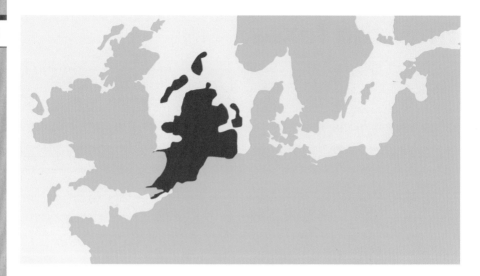

**Above:** The map above depicts part of Northwest Europe about 15,000 years ago, when Britain was joined to Ireland and linked to continental Europe by Doggerland (indicated here by the shaded area).

Fossil evidence suggests that the modern Red Fox has been trotting around Europe for at least 400,000 years, give or take the odd ice age. In Britain its remains have turned up in Wolstonian glacial sediments from Warwickshire, which date it to between 330,000 and 135,000 years ago. The last ice age forced it back south to warmer quarters in southern Europe. However, when the ice retreated north some 15,000 years later, the fox returned. The subsequent flooding of Doggerland – the land mass that linked the UK with continental Europe across the southern North Sea until around 6,500 years ago – separated Britain's Red Foxes from those on continental Europe. Since then the two populations have been unable to mix without human assistance.

Ironically, although North America was the ancestral home of foxes, the *Vulpes* lineage appears to have declined on that continent as it proliferated in Eurasia. Today's Red Fox probably reached North America from Eurasia during two separate glacial periods: the Illinoian glaciation of 300,000 to 130,000 years ago, and the Wisconsinan glaciation of 85,000 to 11,000 years ago. After the latter, two distinct early populations were established, one in Alaska and one in the south. As the ice retreated these populations spread south and north respectively.

# The fox family today

Unfortunately for the non-scientist, it is no longer possible to divide the Canidae family into simply 'foxes' and 'non-foxes'. Recent DNA analysis has revealed that many of the species that bear the common name 'fox' – and which may outwardly resemble our idea of a fox – have different evolutionary affinities. The true foxes, however, are relatively easily defined. These form the genus *Vulpes* (12 species) and include the Red Fox, alongside 11 other species. The other animals that bear the name fox fall into four separate genera, as follows:

**Above:** The Crab-eating Fox of South America belongs in its own genus.

- *Cerdocyon* Crab-eating Fox (South America)

- *Pseudalopex* (or *Lycalopex*) Six South American species, also known as zorros (*zorro* is Spanish for fox)

- *Otocyon* Bat-eared Fox (sub-Saharan Africa)

- *Urocyon* Grey Fox and Island Fox (Americas).

Thus, in evolutionary terms, not every 'fox' is really a fox. In fact, the South American zorros are now thought to be more closely related to wolves and jackals, while the Grey Fox and Island Fox may be the most ancient of all living canids. Casting science aside for one moment, however, all these animals share many of the key characteristics of size, appearance and behaviour that we think of as definitively 'foxy' – which is, of course, how each got its name.

**Below:** A Coyote (left) is larger and more robust than a Red Fox (right).

# What's in a name?

Every living organism has a single, scientific name. This helps prevent confusion when its common name differs from one region to another. Thus, although the Red Fox may also be called Silver Fox, the two names simply represent different colour morphs of the same species and so share the single scientific name *Vulpes vulpes*. A species name is always an italicised binomial (two-part name), with the first word denoting the genus. Sometimes a trinomial (three-part name) is used to denote a race or subspecies – so, for example, *Vulpes vulpes vulpes* is the European race of Red Fox

and *V. v. fulva* is the American race. The full taxonomy of the European Red Fox is as follows.

| | |
|---|---|
| Kingdom | Animalia (animals) |
| Phylum | Chordata (vertebrates) |
| Class | Mammalia (mammals) |
| Order | Carnivora (carnivores) |
| Family | Canidae (dogs) |
| Tribe | Vulpini (foxes) |
| Genus | *Vulpes* (true foxes) |
| Species | *Vulpes vulpes* (Red Fox) |
| Subspecies | *Vulpes vulpes vulpes* (European Red Fox) |

**Right:** The Red Fox is divided into several subspecies. This is the European Red Fox *Vulpes vulpes vulpes*.

# Some other true foxes

- **Fennec Fox** (*Vulpes zerda*) This bijou, rabbit-sized creature is the smallest of all wild dogs, weighing no more than about 1.5kg (3½lb). It also has the most impressive ears: enormous, highly mobile appendages that it uses both to detect rodents and other small prey in its desert environment, and to dissipate the fierce desert heat through a fine network of surface blood vessels. A resident of the Sahara, ranging from Morocco east to Kuwait, this is a strictly nocturnal species that holes up by day in its sand-dune den. Once popular in the pet trade, and also prized by hunters for its lustrous creamy fur, this animal's main natural predators are Eagle Owls.

- **Arctic Fox** (*Vulpes lagopus*) The thick white fur of this medium-sized fox is a telltale adaptation to its life in extreme cold. The most northerly of all land mammals, its circumpolar range extends across the frozen north, from Canada to Scandinavia and Russia. It is roughly the same size as a Red Fox, but has a more compact body, a shorter muzzle and smaller ears, all of which help reduce its surface-area-to-volume ratio and thus minimise heat loss. Other adaptations to its extreme environment include furry pads for walking on ice, and an annual summer moult from a white coat to a brown one, allowing summer camouflage against the snowless tundra. A versatile predator, it eats everything from berries and seabird eggs to lemmings and seal pups, and may scavenge Polar Bear kills. Climate change is causing this species to lose ground to the Red Fox (see box, page 29).

- **Kit Fox** (*Vulpes macrotis*) This small fox inhabits the arid regions of the south-western United States, extending over the border into Mexico. Like its Sahara cousin, the Fennec Fox, it is adapted to surviving the rigours of a desert environment, having a small body, large ears and a largely nocturnal lifestyle. Its grey coat is set off with rusty tones and a black-tipped tail. Prey

**Top:** Fennec Fox.

**Second from top:** Arctic Fox (summer coat).

**Third from top:** Arctic Fox (winter coat).

**Bottom:** Kit Fox.

consists largely of small mammals such as Kangaroo Rats and Cottontail Rabbits, but it will also feed on desert tomatoes and cactus fruits.

The Swift Fox (*V. velox*), found in prairie grasslands a little to the north-east, was once thought to be the same species but is now accepted by most authorities as being a species in its own right.

- **Corsac Fox** (*Vulpes corsac*) This thick-furred little fox is about half the weight of the Red Fox and inhabits the steppes and semi-deserts of central and north-east Asia, from Kazakhstan to Mongolia and Russia. It prefers barren open regions, avoiding mountainous regions and areas of dense vegetation – although its sharp claws make it a proficient tree climber. Food consists primarily of Hamsters, Ground Squirrels and other small mammals, although it will take fruits when required. Nomadic by nature, this is one of the few foxes that may sometimes form packs. Its silky winter coat is in great demand for the fur trade, while natural predators include Grey Wolves, eagles and Eagle Owls.

**Top:** Swift Fox.
**Second from top:** Corsac Fox.
**Third from top:** Bat-eared Fox.

- **Bat-eared Fox** (*Otocyon megalotis*) This endearing, domestic-cat-sized animal inhabits sub-Saharan African savannas, where it has been around since the middle of the Pleistocene epoch, some 800,000 years ago. It belongs in a genus of its own, and is easily distinguished from other canids by its unique combination of huge ears and dark, Raccoon-like face mask. This is the only dog to subsist largely on insects – mostly termites – for which it forages in small family groups, radar ears cocked to the ground for the rustle of its prey beneath. Being on the menu for many larger predators, Bat-eared Foxes are equally vigilant for danger from above and are quick to dash for cover. The litter of 4–6 cubs, born in an underground den, is guarded by the male while the female is out foraging.

# Arctic invasion

There is no love lost between Red Foxes and Arctic Foxes. The two species are natural competitors and the Red Fox, being the larger, faster and more powerful, will drive out its northern relative wherever the two come across one another. Indeed, in experiments in which both species were released from fur farms onto islands in the Russian Arctic, the Red Foxes killed their rivals outright. Elsewhere Red Foxes have been observed digging Arctic Foxes out of their burrows, killing them, and in some cases eating them. Due to this natural antipathy between the species, the Red Fox effectively sets the southern limit for the geographical distribution of the Arctic Fox.

The antagonism between the two species need not be a problem. The two animals have evolved naturally to fill different niches. The Arctic Fox, with its white coat, more compact build and various behavioural adaptations, is an incredibly hardy animal perfectly suited to survive in the frozen Arctic – an extreme environment that the Red Fox cannot handle. The Red Fox, by contrast, has evolved to survive as far north as the boreal forests, but no further. In theory, then, never the twain should meet. In practice, however, the two species are bumping into one another with increasing frequency – and with dire consequences for the Arctic Fox. Most scientists believe the reason for this is climate change: global warming has seen a drastic reduction of the Arctic pack ice and a 'greening' of the tundra as the boreal forests spread north. The Red Fox is seizing this opportunity to expand its range northwards, colonising terrain across parts of Russia, Canada and Alaska that was once exclusively the domain of the Arctic Fox. The IUCN now lists the Arctic Fox as one of the 10 animals worldwide most vulnerable to climate change as a result of changes in its tundra habitat. Where the Arctic Fox is the loser, the Red Fox is the winner.

**Above:** Arctic Fox in its natural habitat.

**Above:** Red Fox in Arctic habitat.

**Below:** A young Polar Bear chases an Arctic Fox in the Arctic National Wildlife Refuge, Alaska.

# Red Foxes around the world

The Red Fox has adapted easily to the many different environments it has colonised around the world, adapting its form and behaviour in the process. Some scientists have proposed at least 45 subspecies of Red Fox, reflecting differences in size, skeleton, teeth, colour and other physical properties. This remains a contentious issue, as other scientists contend that many proposed subspecies are not sufficiently distinct to warrant these divisions, and in turn propose no more than five. Either way, the fact remains that Red Foxes in one part of the world can look very different from those in another.

In general, the different subspecies (or at least forms) of the Red Fox fall into two groups: the northern races, found across Europe, northern Asia and North America, which are large, with brightly coloured coats; and the southern desert races, found in the Middle East, North Africa and southern Asia, which are halfway in size between Red Foxes and smaller fox species, with duller coats, and proportionally larger ears and longer legs.

These southern desert Red Foxes also behave in a number of different ways from their more northern cousins, including living in larger, non-territorial groups.

**Below:** Red Foxes in desert environments can be very pale in colour.

Red Foxes in central Asia are intermediate between the southern and northern races.

**Above:** Northern Red Foxes are larger and more brightly coloured than their southern counterparts.

The cross-breeding of different subspecies, introduced by people across regions, has muddied the genetic waters somewhat. Thus British Foxes have cross-bred with those imported from Germany, France, Belgium and possibly Scandinavia and Siberia. European Foxes introduced to eastern parts of the USA in the 18th century have cross-bred with native North American populations. Within North America, introduced Foxes of the eastern races may have been cross-breeding with those in California of a separate race.

The status of the North American Red Fox is especially complicated. Historically, scientists classified it as a completely different species (*Vulpes fulva*) from the Eurasian Red Fox. Today it is widely accepted that the American Red Fox (*Vulpes vulpes fulva*), sometimes called the Eastern American Red Fox, is simply a North American subspecies. However, within North America there are two major 'groupings' (or clades): those in Alaska and western Canada are grouped together with Eurasian Red Foxes to form a Holarctic clade; those in southern and eastern states form a genetically distinct

**Above:** A dark-coated Red Fox in Alaska.

**Below:** This Red Fox, caught on camera trap in Iran's Dar-e Anjir Wildlife Refuge, shows the pale colouration typical of arid regions.

Nearctic clade. This Nearctic clade itself divided during the Winsconsinan glaciation. When the ice melted around 10,000 years ago, Foxes in the east (boreal subclade) followed the boreal forests north up to eastern and central Canada, while those in the west (mountain subclade) colonised the alpine and subalpine habitats of the Cascades, Sierra Nevadas, the central Rocky Mountains and other mountain regions. Claims that all Foxes along America's eastern seaboard are largely descended from European Foxes, via historical translocation, have been disproved by DNA research, proving that most of these animals have arrived via natural range expansions and are therefore indigenous and not invasive.

The Red Fox is a vital part of the ecology in all parts of the world to which it is native. As an alien invader, however, this resourceful predator has proved disastrous – as Australians have discovered. The Red Fox was originally introduced to Australia in the Melbourne area during the mid-19th century to provide sport (Fox-hunting) for homesick expat Brits. By the early 20th century it had colonised South Australia, New South Wales, Queensland and Western Australia, spreading at rates of up to 160km (100 miles) per year – in some cases with further help from humans. Today the Red Fox roams the entire country except for the northernmost arid and tropical regions, penetrating deep into the hot interior and colonising at least 18 offshore islands.

This invasion has had a devastating impact on Australia's native wildlife, which has not evolved to deal with a predator of this nature. Since its introduction, the Red Fox has been directly implicated in the extinction of the Desert Rat-kangaroo, and the severe decline of numerous other small marsupials, including the Brush-tailed Bettong, Rufous Bettong, Eastern Barred Bandicoot, Bilby, Numbat and Quokka.

Many of these species now live only on islands or in other restricted locations where Foxes are absent or rare. Foxes have even had an impact on the much larger Eastern Grey Kangaroo, in one region taking up to 35 per cent of juveniles during their first year. Other animals affected range from birds such as the Little Penguin and Malleefowl, to reptiles like the Bearded Dragon and Sand Goanna.

This menace has been hard to fight. Eradication has proved impossible, but a combination of poisoned baiting and hunting (legal in all states), with the encouragement offered by state bounties, has offered effective management in some areas, generally at a local rather than a landscape level. It has also been suggested that the reintroduction of Dingos to areas where they were once native has helped to both reduce Fox numbers and encourage the return of native wildlife. Worryingly, the 21st century has seen the first records of Red Foxes in Tasmania, after deliberate illegal introductions in 1999, and their impact on that island – especially rich in native marsupials – remains to be seen. There are currently no Foxes in New Zealand, and it has been illegal to import them there since 1867.

**Above:** The Eastern Barred Bandicoot is one of numerous small marsupial species that has declined dramatically owing to predation by the invasive Red Fox.

**Above:** A Rosenberg's Goanna on Kangaroo Island, South Australia. This large reptile is the largest predator here, in the absence of Foxes, but is declining on mainland Australia where it cannot compete with the introduced canine.

# Grabbing a Bite

**Food is a critical resource to any animal, determining where it can live, how much space it needs and when it is able to breed. There are few animals better equipped at finding food than the Red Fox. Its amazingly varied diet, and the impressive repertoire of strategies it uses to satisfy its appetite, allows this most unfussy of eaters to live pretty much anywhere. Though evolved primarily as a specialist predator on small mammals, it is very much an omnivore and will happily scoff anything from fruits to carrion, adapting to habitat, season and circumstance with the ease of a true opportunist.**

A Red Fox requires around 450–550g (1–1¼lb) of food a day – roughly 10 per cent of its body weight – in order to generate the heat and energy it needs to survive. Adults spend on average about one-third of their waking hours looking for food. They spend even more time doing so during late spring and early summer, when they are feeding pups, and also in environments where food is scarce, such as those on mountains and in deserts.

The Red Fox has the most catholic tastes of any wild dog, with its menu comprising everything from rabbits and rodents to earthworms, eggs, fruits including berries, carrion and human waste. Research has produced some impressive statistics. One study in the former Soviet Union identified 300 animal species within the diets of local Red Foxes. It is not always easy to identify exactly what Foxes eat and where they get their food. The remains of an animal in a Fox's scat (droppings) is not necessarily proof that the Fox actively hunted and killed this prey; it may have come across it as carrion. Equally, some foods that break down entirely during digestion may not appear in scats at all.

**Opposite:** Rabbits make up an important part of the Red Fox diet in many regions.

**Below:** An adult Red Fox spends about one-third of its waking hours searching for food.

**Above:** The Lion is an obligate carnivoran: it depends upon a diet of almost exclusively meat.

**Below:** The Fox diet includes plenty of fruit – especially during autumn.

The Red Fox is what is known as a facultative carnivoran. Carnivorans are members of the order Carnivora (see page 21), whose diet comprises at least 60 per cent meat – as opposed to those, such as some bears, that consume a larger proportion of other foods. It is not an obligate carnivoran – a carnivore such as a Lion that eats almost exclusively meat – but given the choice it does prefer meat over other foodstuffs. Indeed its system is adapted for this: a 1987 University of Chicago study discovered that a Fox's digestion rate for mice (89 per cent) is much higher than that for fruits (51 per cent).

Foxes adjust their diets according to the seasonal abundance and availability of their food. Many of the small mammals on which they prey, such as voles, are subject to boom-and-bust population fluctuations, and Foxes gorge themselves in times of plenty – over 60 voles have been found in the stomach of a single Fox – then find alternatives at other times. Thus fruits are popular in autumn and insects in summer. Like all predators, Foxes are subject to what is known as the optimal foraging theory: in other words, their feeding behaviour constantly changes to ensure that they receive the greatest energy returns for the expenditure of effort.

# Furry foods

**Above:** Voles, such as this Short-tailed Vole, are the preferred prey of foxes in many areas.

The Red Fox is a specialist hunter of small mammals of generally no more than 3.5kg (7¾lb) in weight. Top of the menu are rodents, which make up some 50 per cent of its diet in rural areas. Where available, Foxes seem to prefer voles (Microtinae subfamily) to rats and mice (Muridae), with the Short-tailed Vole being a particular favourite in the UK. Elsewhere around the world the rodent diet may extend to Ground Squirrels, Pocket Gophers, Hamsters and Gerbils, depending on what is locally available.

**Below:** Down in one gulp: it takes a lot of small rodents to satisfy a hungry fox.

**Above:** Foxes seldom eat Moles, but adults will offer them to cubs as playthings.

**Above:** Foxes seldom capture animals as large as this fawn, but will feed on any carcasses they find.

**Below:** A Hedgehog's armoury of spines provides a prickly challenge for a hungry Fox.

Some small mammals – notably Shrews and Moles, which produce an oily, musky secretion – are less appetising to Foxes than others. These may be cached for consumption at a later date, or caught alive and taken to the cubs as playthings. Foxes do eat Hedgehogs, although it is hard to work out how much of this diet comes their way via road kill. It is perhaps telling that the Hedgehog population in Bristol increased significantly during the mid-1990s, when an outbreak of mange caused the Fox population to plummet.

Sitting a little heavier on the mammal scales are lagomorphs – Rabbits and Hares – which are significant prey for Foxes in many regions. Numerous other small mammal species have been recorded in the Fox's diet around the world, from porcupine and possum to Raccoon, Otter, Stoat and, in Australia, wallaby, although in many cases the animals may well have been consumed as carrion. The largest animals actively predated by Red Foxes include the young of ungulates such as deer and Wild Boar. The Fox's ability to capture such large prey – if the mother is absent, or circumstances otherwise allow – confirms that it is capable of killing newborn lambs. Such incidents, however, happen only rarely.

On a more macabre note, Foxes have been known to feed on human flesh. Rare cases include the digging up of infants from pauper's graves in London, and the partial consumption of a suicide victim in the Australian Outback. They will also, in extremis, resort to cannibalism and feed on their own kind. Foxes are undeterred by our ethical qualms. Life is a daily battle to fulfil their basic energy needs, and food is, after all, food.

# Also on the menu

**Above:** A Fox may capture birds as large as a swan.

Foxes will happily turn their predatory attention from fur to feathers. Birds make up a relatively small proportion of the Fox menu in most areas, but where circumstances allow a Fox will capture anything from small songbirds such as thrushes to larger fare such as game birds, pigeons and waterfowl. Birds are most easily grabbed when they are on the nest, especially at night and during windy conditions, when flying is difficult. They generally form a larger part of the diet during spring and early summer, when adults bring them back for cubs at the den. Lead shot found in Fox scat suggests that these arch opportunists sometimes capture birds that have been winged by hunters.

Birds' eggs are a particular delicacy. Foxes will carry them carefully back to their cubs and often cache them for later consumption. The eggs of other animals are equally welcome: studies from as far afield as Turkey and North Carolina have revealed that Foxes are especially fond of turtle eggs, with 89 per cent of Loggerhead Turtle nests in one study in Turkey having been raided by Foxes. Lizards and other reptiles are important prey for Fox populations in some arid regions, such as the north Caucasus steppes of Russia. Frogs also make it to the menu, as do fish, where available. Foxes have been observed using their forepaws to hook goldfish out of garden ponds, as cats do, and even dragging up a fisherman's line to steal the catch at the end.

**Below:** A fish makes a rare but welcome treat for any Fox.

**Above:** A clutch of Pheasant eggs makes a tasty treat for any Fox.

**Above:** Worms are a very important part of the Fox diet.

**Below:** A Fox tries to wrap its jaws around an apple.

Invertebrates make up a decent portion of the Fox's diet. Insects, including beetles, moths, crickets and craneflies, are all fair game; in some desert regions they are a vital source of nutrition. The creepy-crawlies of which Foxes are most fond on UK shores are earthworms, especially during late summer when these can provide up to 60 per cent of a Fox's calorific intake. Worms are important to growing cubs, as they are about the only prey that they can capture alone in the first few weeks after they stop suckling from their mothers. Indeed, studies in the UK have shown that cubs growing up in areas with high rainfall, where worms are easier to come by at the surface, tend to grow significantly larger than those in less worm-rich areas.

Although Foxes may officially be filed under carnivores, they are not averse to a bit of greenery. Top of the plant menu come energy-rich foods such as fruits and berries. In late summer and autumn, when blackberries and blueberries are fruiting, these tasty treats may dominate a Fox's diet. Other popular fruits include rose hips, cherries, raspberries, apples, plums, grapes and acorns, and yew berries – toxic to us – are guzzled with impunity. Foxes also eat a variety of less juicy vegetable matter, from grasses, sedges and cereal crops such as corn and barley, to tubers and even hazelnuts.

# The importance of leftovers

A Fox's powerful nose equips it well for sniffing out dead things, and carrion is indeed an important source of food. In areas that Foxes share with big predators, such as Grey Wolves, lynxes and hyenas, they are adept at locating the larger animals' kills and stealing leftovers. Elsewhere they may feed on road kill or retrieve animals shot by hunters and not yet claimed. A large carcass may draw a larger than usual gathering: during the hard winter of 1947, 12 Foxes were observed scavenging together on a Red Deer carcass in the Scottish Highlands. Foxes tend only to visit carcasses at dusk and dawn, and they fare less well in windy conditions, when the scent is dispersed, than in calm ones.

**Above:** A Fox scavenges from carrion.

Human leftovers – our discarded food and other refuse – help sustain Foxes in both rural and urban areas. In the latter they may constitute up to 80 per cent of a Fox's diet. Foxes are not picky: while they prefer meat and bones, they tuck into anything from rotten fruits to bread and potato peelings. Some of this is scavenged from bins, skips and discarded takeaways, but in the UK plenty of it also comes as deliberate handouts from householders, many of whom enjoy having Foxes around and go out of their way to attract them. Indeed, Foxes raid bins less frequently than is commonly assumed, and the design of modern wheelie bins tends to thwart their ambitions in this area.

Landfill sites offer rich pickings for Foxes, as was amply demonstrated in BBC's *Autumnwatch*, which documented the fortunes of a community of Foxes around Pitsea Landfill site in south Essex. Numerous Foxes set up territories in the area, subsisting almost entirely on the daily deliveries of rubbish that were dumped on the 680-acre (275ha) site.

**Below:** Landfill sites provide rich pickings for Foxes.

## Farmyard felon

Any child raised on Beatrix Potter or Ladybird books will believe that Foxes have an insatiable appetite for farmyard hens and stop at nothing to secure their prize. Certainly, Foxes have, over the years, gained notoriety for their raids on livestock – from chickens to lambs, piglets and even, in the New Forest, foals, not to mention the legions of beloved pet rabbits and guinea pigs ripped unceremoniously from their hutches. The reality, however, does not justify the hysteria. Foxes do indeed take chickens where they can, but it is estimated that they are responsible for no more than 2 per cent of losses on poultry farms. They may even, on very rare occasions, target newborn lambs.

Before rushing to judgement, however, we should remember that a Fox is not necessarily responsible for killing the animal on which it is seen feeding. Livestock dies of natural causes, and a far greater danger to many domestic

and farm animals are stray dogs, of which some 121,000 were collected in the UK in 2011 alone. Meanwhile, other wild predators of small livestock include Mink, Stoats and birds of prey. Of course, no predator will turn down an easy meal – Foxes were responsible for the demise of 11 penguins at London Zoo in 2009. However, a combination of decent enclosures and good animal husbandry keeps most livestock and pets safe from Foxes.

**Above:** The relationship between Red Fox and farmyard chicken is not always a happy one.

# A view to a kill

**Below:** A Fox does much of its hunting after dark.

Foxes find most of their food between sunset and sunrise. They patrol at least a part of their territory every night, over time building up a detailed picture of where food is most likely to be, and constantly refining this as conditions change. While hunting and foraging they are always on the move, trotting at an average speed of 6–13kph (4–8mph) but capable of bursting into a 50kph (31mph) sprint when required. Few obstacles stand in their way: they are good swimmers and proficient diggers, and can scramble over a 2m (6½ft) garden fence with ease. Catching live prey requires skill and cunning, both qualities with which the Fox is amply endowed. Nonetheless, success is by no means guaranteed.

The Fox's best-known – and certainly most spectacular – hunting technique is the cat-like 'mousing pounce' that

it uses to capture small mammals. This is a leap into the air from a standing position, to land on the unsuspecting prey and snap it up before it has a chance to escape. The technique is perfected from about six weeks of age, when cubs first practise on beetles and other insects. The prey, which may be hidden beneath grass or snow, is first detected by sound. The Fox moves slowly, ears cocked, head swinging from side to side, until it picks up a telltale rustle. It then freezes until it can locate the exact source of the sound. Prey pinpointed, it rears onto its hind legs, bends its knees and launches itself into the air – usually at an angle of 40 degrees – before landing, front-feet first, on its prey. While in mid-air it steers with its tail.

A mousing pounce, which may cover 5m (16½ft) in length and 2m (6½ft) in height, demonstrates not only a Fox's agility, but also how specialised the species is in capturing small rodents. The higher it leaps, the harder it comes down, which helps when breaking through a crust of snow to capture a mammal scurrying about a metre or more below. Strikes are impressively accurate, and there is some evidence that Foxes make use of the Earth's magnetic field to orientate themselves. The disadvantage of targeting unseen prey in this way, however, is that the Fox does not know

**Above:** The cat-like 'mousing pounce' serves to capture small rodents in long grass.

**Below:** Keen hearing helps to detect prey moving beneath a layer of snow.

what it is getting and may capture something less desirable, such as a Shrew, which it may then discard.

For larger prey, such as Rabbits, Foxes stalk as close as possible, belly to the ground and using every available bit of cover, before rushing their quarry in one lightning-fast dash, hoping to seize it before it disappears into the safety of its burrow. This again, in many ways, is more cat-like than dog-like behaviour. Where no cover is available, for example in an open field, a Fox may simply approach its prey in full view, freezing every time the quarry looks up, then continuing to move closer when it looks down until it gets close enough for the final rush.

Foxes study the behaviour of their prey, and learn from experience how to outwit it. In the vicinity of a Rabbit burrow, for example, they make themselves obvious to the sentinel Rabbits – which keep an eye out for danger while the others graze – then retreat to the cover of a hedge, feigning lack of interest. Once the sentinels relax their guard they launch the attack. A Fox hunting Arctic Ground Squirrels in Canada was observed watching the rodents as they dashed into their burrow after each attack then re-emerged at a rear entrance to check whether the coast was clear. After this happened several times, the Fox moved immediately to the rear entrance, following its first charge, just in time to snatch its unsuspecting quarry as soon as it appeared.

**Above:** A Fox disappears at speed to safeguard its catch.

**Below:** A Ground Squirrel makes a substantial meal for this North American Red Fox.

# The cunning, wily fox?

Few animals have generated more myths than the Fox, which features in the literature and folklore of numerous cultures – generally in the role of cunning trickster (see page 105). These myths often extend to the way in which Foxes are thought to hunt, and it can sometimes be hard to separate the facts from the fiction. Beliefs popular in British rural folklore include:

- A Fox only hunts away from home – that is, it would never attract attention to itself by, say, raiding a chicken coop in the immediate vicinity of its earth.
- Foxes play dead to lure prey close enough to catch.
- Foxes urinate on Hedgehogs to get them to uncurl.

While none of these beliefs has been satisfactorily proven to science, there is plenty of anecdotal evidence to suggest that Foxes are capable of pulling some surprising stunts. Hunters in America, for example, have observed a Fox playing with a stick close to a group of ducks, repeatedly picking it up and dropping it. Once the Fox retreated the curious ducks waddled over to the stick for a closer look, whereupon the Fox sprang back and grabbed one. A YouTube video also includes extraordinary footage of a Fox trying to pull in a fisherman's line, on the end of which is a huge catfish.

**Above:** A Red Fox makes away with somebody else's prize salmon.

A Fox usually kills mammals and birds with a bite to the back of the neck that severs the cervical vertebrae. It can also deploy a surprising subtlety when more delicate items are on the menu. When hunting worms, for example, it grips the slithery morsel in its incisors, but instead of tugging violently, which would break the worm, it pauses until the worm relaxes, then pulls gently and steadily so as to extract the whole animal, unbroken, from the ground. Parents teach their cubs this same technique by example.

When picking fruits, a Fox may stand on its hind legs to reach the juiciest prizes, and pluck blackberries from the plant with such delicacy that, reputedly, not a leaf is disturbed. Foxes also play with food that they do not eat, such as Moles or Shrews, or bring it back alive to their cubs for a little hunting practice – very much in the manner of cats.

**Below:** Foxes, like cats, often play with their prey.

# Cache as cache can

**Above:** A Red Fox in the Vosges, France, with a mouthful of voles.

Foxes are very possessive about their food, defending it fiercely from other predators and rivals, including more dominant Foxes. They generally eat the choicest items on the spot. Other, less tasty parts, or any food they cannot finish, is often take away and hidden, to be retrieved a day or so later. This behaviour is known as caching. A Fox will bury its cache under leaves, soil or snow, using its front paws to dig, then rake back a covering layer to conceal it. It will then depart carefully, often moving away backwards and obscuring its tracks as it does so.

Foxes use what are known as scatter caches – in other words, they bury their treasure in a number of places over a wide area, as opposed to selecting just one place (a larder cache). This means that they are less vulnerable to losing everything if a competitor sniffs out their hoard – but it also means that they have to rely on memory to retrieve their goodies. Experiments in which Foxes have failed to find other items buried within a few metres of their caches suggest that it is memory rather than smell that leads them back to a hidden location. Observers have also seen Foxes urinate on caches that they have emptied, which suggests that they might be using scent to identify sites to which they need not return.

# Slaughterhouse or supermarket?

A charge often levelled at Foxes is that of surplus killing – that they kill more prey than they eat. A typical story tells of a Fox that enters a chicken coop, slaughters a dozen chickens in a frenzy of feathers and exits with only one of them. It is not only chickens that this applies to: Foxes can do the same with a hutch full of Rabbits, a pen full of game birds or – in the wild – a nesting colony of gulls, on a wild, windy night. This is an emotive topic, with all its connotations of cruelty and waste. Suffice it to say, a highly evolved hunter such as a Fox, which is dependent for its own survival upon killing, has no concept of cruelty – nor indeed any consideration for the welfare of its victims. To suggest otherwise is sheer anthropomorphism.

A Fox is hardwired to take food whenever and wherever it can. It has no knowledge of what opportunities may or may not await tomorrow, so must fill its boots when the chance presents itself. Prey trapped in a single place, such as a chicken coop, is, to a Fox, rather like a stacked supermarket shelf: just as we do our weekly shop to avoid the wasted effort of multiple trips, a Fox takes as much as it can when it can. The fact that it may leave the scene without removing the evidence is most likely due to the fact that it has been disturbed. It is hard for a Fox to carry more than one chicken at a time, and the hole through which it has entered may be too small for anything bigger. Experiments have shown that, if left undisturbed, a Fox will return to a hen coop in which it has left behind dead chickens and remove them one by one – some to eat immediately, some to take to its cubs and some to cache for later.

**Below:** This chicken has died for food – not pleasure.

# Passing It On

Crucial to the success of the Red Fox worldwide has been its unfailing ability to produce an endless stream of baby Foxes. The species is, in other words, an excellent breeder, with a robust and efficient reproductive strategy that allows populations to recover quickly from any setback due to disease, persecution or other threats. This strategy involves a number of important strands, from careful mate selection to vigorous mating and solicitous parenting.

## Getting in the mood

Studies of both wild and captive Red Foxes have taught us a great deal about how the species goes about producing the next generation but – as ever, when it comes to sex – much of the topic still remains shrouded in mystery. For adult Foxes the annual breeding cycle begins in late summer, when the year's cubs have grown big enough to make their own way in the world. Both males and females, freed from the rigours of parenting, begin to return to breeding condition. This new phase is reflected in their reproductive organs, with both the vixen's ovaries and the dog's testes swelling in size. The dog starts producing sperm once again after the summer's hiatus.

**Opposite:** A vixen nursing her litter.

**Below:** A moment of intimacy between a courting pair.

**Above:** Courtship behaviour between a male and female Fox involves a complexity of body language and vocalisations, both tender and aggressive, with the male not letting the female out of his sight.

The breeding window is a brief one. Red Fox vixens come into heat once a year, with their ovulation triggered by shortening day length rather than by any signs of enthusiasm in their mates. This period of heat, known as oestrus, generally lasts for about three weeks, during which there is a critical phase of around three days when the vixen is receptive to fertilisation. Older females tend to come into oestrus a little earlier than younger ones. It is now that a dog Fox must make his move if he is to sire the next generation.

In the UK and across much of the northern hemisphere, the peak mating season is January. As this period approaches, the female becomes increasing vocal, advertising her presence to suitors with a distinctive scream. As his window of opportunity approaches the dog Fox follows the vixen day and night, not daring to let her out of his sight for fear of missing his moment or losing her to a rival. His unremitting attentiveness at this time is known as mate-guarding, and he and the vixen may sleep, travel and even hunt together.

# Getting down to it

During her three days or so of receptiveness the vixen scent marks frantically around her territory, spreading a pungent message that she is ready for action. The dog follows close behind, checking out all the vixen's scent marks for the chemical signal that indicates her readiness, scraping into them with his paws and urinating on everything with renewed vigour to reinforce the territorial claim. For these few days he hardly feeds.

Once the vixen feels ready to mate, she entices the dog with submissive posturing, rolling over before him to expose her underside, pushing her rear end towards his face and generally – if we are to resort to human stereotypes – teasing and flirting. The dog's first attempt to climb up, signalled by raising a forepaw onto her back, is generally rebuffed aggressively. This may happen several times before the vixen finally allows him to mount her, using both forepaws to keep a firm grip. Things become very noisy as the mating ritual intensifies, with the vixen keeping up a constant chorus of short wails, shrieks and chittering sounds of invitation.

**Below:** A dog Fox uses his forepaws to grip the vixen when mounting her.

The mating itself is a painful and protracted affair. The dog Fox – like most mammals, except humans – has a penis bone called a baculum, detached from the rest of the skeleton, which serves to maintain his erection during mating. Once the dog has entered the vixen, a bulb-like tissue mass at the tip of the baculum, called the bulbis glandis, engorges with blood and the female's vagina simultaneously contracts. This is undoubtedly painful for the vixen, who issues shrill screams. What is more, it causes the mating pair to become locked together until the swelling subsides.

This locking process, known as a copulatory tie, may last for just a couple of minutes or for more than an hour. Once joined in this way, the Foxes rearrange themselves to stand back to back, the dog lifting one hind leg over the vixen's rump, and can thus better defend themselves should they need to. It may seem bizarre and rather grotesque, but the copulatory tie gives the sperm the best odds of fertilisation. It also prevents other males from mounting the female, thus ensuring that the dog's sperm gets first shot. Studies have shown that a copulatory tie takes place only after a successful ejaculation.

**Below:** Caught in the act: a pair of Foxes locked in a copulatory tie.

# To have and to hold?

Once mating is over, a mated pair of Foxes remains together, the two often curling up to sleep side by side. However, before becoming dewy-eyed about the touching beauty of canine love, we should perhaps look a little closer at the complexity of their relationships.

Foxes can be described as socially monogamous. This means that a resident pair remains within a permanent territory and comes together every breeding season to rear a litter of cubs. This bond may last for life – unless one partner is killed, in which case the other pairs up elsewhere, possibly moving on in the process. Studies and anecdotal evidence have suggested that such bonds can generate strong emotional attachments. A dog Fox, recently bereaved, has been observed wandering the shared territory for some days calling plaintively as though in mourning for his missing partner.

Nonetheless, it is not quite so simple as 'till death us do part'. While a dominant male may pair up for life with a dominant female, neither partner is averse to a bit on the side. When the vixen is out of oestrus, the dog may seek out other females to mate with – even temporarily expanding

**Above:** A dog Fox mate-guarding a vixen.

## Breeding rights

Not all vixens get to breed. Many factors have a bearing on breeding success. Food supply is of critical importance. Studies in Sweden have shown that the ovulation rate of females is tied to the abundance of voles, with a bumper year promoting their fertility. Studies in Russia, conversely, reveal that females do not come into oestrus at all during poor food years.

Social status is also important. Thus in an area populated by several vixens it may be only the dominant female that gets to breed. Her aggression and monopolising of food may suppress reproduction in her subordinates. Stressed vixens undergo a hormonal change that inhibits the production of progesterone – the hormone that prepares the lining of the uterus to receive the egg – and thus, if harassed by a dominant individual in this way, subordinates may abort the fertilised egg before implantation. Breeding success also increases with age, which is often tied to social status. Fewer than half of Foxes breed successfully in their first year. Diseases such as sarcoptic mange (see page 89), also do not help.

**Right:** A dominant female chases away her subordinate.

**Above:** Sticking together helps provide territorial security to a pair of Foxes, but does not necessarily prevent a little extra-curricular mating.

his range to do so – in order to maximise the spread of his genes. Mortality among dog Foxes runs high during this period. Many are too fixated on mating to feed, and thus lose weight and condition, while others may be killed in fights or get run over during their wanderings. Vixens, meanwhile, may mate with more one partner, accepting the advances of other suitors that enter their territory when the dominant Fox is not around to drive them away.

This promiscuity has its advantages. Males that mate with more than one female (polygyny) are, in doing so, spreading their genes around and so increasing the chances of more of their progeny surviving. Females that mate with more than one male (polyandry) are ensuring that they receive a broad mix of genetic material, thus upping the chances for at least some of their progeny. Against this, however, must be balanced the advantages of staying with a single partner, who enjoys territorial dominance and can therefore provide food and help defend the cubs.

Ultimately there are no inviolable laws when it comes to the Fox mating game. Behaviour varies with habitat and the availability of food. In areas supporting a high density of Foxes, extra-pair matings are more common. Studies of urban Foxes have shown that a vixen's litter often contains pups of varied paternity. Conversely, where food is scarce, monogamy tends to be the norm. Either way, it seems that dominant vixens never produce litters with subordinate males, and that a dominant male continues to attend his mated vixen, even if some of her cubs are not his.

# Pregnancy and birth

A Red Fox vixen undergoes a gestation period of 49–58 days, typically 52 days (or roughly 7.5 weeks), which is the shortest known of any canine and compares with 58–65 days for domestic dogs. Implantation – the lodging of the fertilised egg in the wall of the uterus – takes place around 12 days after a successful mating, when the vixen's progesterone production is at its peak.

Pregnancy is not immediately obvious in a vixen, as her size and shape change very little. Generally it is her behaviour that gives the game away. A pregnant vixen's appetite will increase considerably. Meanwhile, she will start searching for and preparing an earth, ready for the birth. In the UK this process takes place during February.

Earths or dens generally consist of a tunnel dug under the ground leading to a small, hollowed-out chamber, in which the vixen has her cubs. Typical denning sites in urban and suburban areas are under garden sheds or gravestones in churchyards, although many other sites are chosen – including even under the floorboards of an

**Above:** A pregnant vixen near her earth in a graveyard.

**Left:** The size increase in a pregnant vixen is not always apparent until near the end of her gestation.

occupied house. In the countryside a Fox will dig its own earth, or make use of an old Rabbit warren or Badger sett, sometimes sharing the latter with the Badgers, if it is large enough. As the birth date approaches, the vixen spends most of her time denning down, relying on the male to bring her food. By now her teats have become bare and, two weeks before the birth, her mammary glands emerge.

The baby Foxes – known as cubs, kits or pups – are born in early spring, typically mid- to late March in the UK. The vixen generally gives birth underground, in the breeding chamber at the back of the earth. The average litter size is 4–6 cubs. Eight is the largest known for a UK vixen, although there is an unconfirmed record of 13 from Tippecanoe County, Indiana, USA. Litter size tends to correlate with the size of the vixen and the quality of the habitat: the larger the Fox and the more food there is available, the bigger the litter. However, as vixens typically have just three pairs of teats, it is difficult for them to raise a litter of more than six cubs.

**Below:** A vixen suckles her newborn litter in the earth.

A Fox burrow is known as a den or earth. This underground retreat comes into its own during the breeding season, when a vixen requires somewhere secure in which to give birth and look after her cubs during their first few vulnerable weeks. Red Foxes prefer to dig their earth on well-drained soils – often on a slope or hillside, or among tree roots. Favourite spots in urban areas include railway embankments and spaces under garden sheds. Although able to do the spade work themselves – digging initially with the front paws, then kicking back the earth with the hind legs – Foxes often take over the burrows of other large animals, including Rabbits and Badgers or, in North America, Woodchucks and Porcupines. In some cases they may cohabit with their neighbours, the two species using separate chambers within the same burrow complex.

A typical earth consists of a tunnel that leads down at roughly 45 degrees and broadens out into a den, with numerous side branches leading off it. The main passage extends inwards, on average some 5–7m (16–23ft) – occasionally much longer – and descends 0.5–2.5m (1½–8ft) below ground. A good den may last for decades, offering a home to generations of Foxes. Each spring the adults clear out any excess soil or debris, scattering it in a telltale spoil heap outside the entrance. This area becomes trampled down when it later serves as a playground for the kits. Outside the breeding season Foxes have less use for a den and prefer to lie up above ground in thickets or other cover – although they may retreat to the den during bad weather. Foxes living in desert areas spend more time in a den by day to escape the scorching sun.

**Above:** A Fox uses its front paws to excavate an earth.

**Right:** Badger setts, sometimes used by Foxes, are easily identified by the well-worn tracks around them.

**Below:** Graveyards provide excellent earth sites for both Badgers and Foxes in suburban areas.

# New arrivals

Fox cubs are born with the eyes and ears closed, and a pink nose that turns black within a week. The average weight is 100g (3½oz) – about that of a small lemon – although it may vary from 50 to 150g (1¾–5¼oz), and the tiny head measures on average 41mm (1½in). The cubs are covered in a fine layer of woolly grey fur, with a white tail-tip showing even at birth. This fur does not provide enough insulation for them to regulate their own body heat, and for the first 2–3 weeks they are dependent upon the body warmth of their mother, who huddles close to them like a living duvet to prevent hypothermia.

For the first 2–3 days the vixen does not leave her new charges at all – not even to drink. She is dependent upon the dog Fox to bring her food, although she does not allow him into the earth, requiring him to leave his offerings at the entrance. Indeed, the male Fox probably does not lay eyes on his cubs (if indeed they are all his) until they first leave the earth at about four weeks. If he

**Below:** By two weeks, cubs' eyes and ears are open and their fur has turned a chocolate brown.

is late with his food, the vixen goes to the mouth of the earth and barks sharply to make her impatience known.

At 10–14 days the cubs' eyes and ears open and the first milk teeth appear in the upper jaw, followed a couple of days later by those in the lower jaw. By two weeks their fur has turned a chocolate-brown; at three weeks it develops black smudges around the eyes, and at six weeks it has assumed its adult colour, although is still woolly in texture. It is not until eight weeks that the coat acquires the long guard hairs that give it its glossy sheen. The eyes also undergo a change of colour: a pale cloudy blue when first open, due to the eumelanin pigment in the iris they turn deep amber at about 4–5 weeks, once the lipochrome pigment is produced.

Down in the den the cubs are playing, napping and grooming. For the first 4–5 weeks they depend completely upon their mother's milk, suckling regularly. She produces some 300ml (½ pint) of milk per day, increasing to 800ml (1½ pints) by the time her litter is 16 days old. The milk is high in sugar and protein but relatively low in fat, enabling the cubs to build muscle quickly. One study

**Below:** After five weeks, the cubs venture above ground, but continue to stick close to their mother and suckle regularly.

**Above:** Cubs continue to suckle for 6–7 weeks.

of growing Fox litters in Norway concluded that about 30g (1oz) of milk per 100g (3½oz) of cub is required for healthy development.

At this stage the cubs cannot evacuate their own bowels without stimulation, so their mother regularly licks their perineal region and eats the waste that emerges. Distasteful as this may seem to us, it is part of a female Fox's meticulous cleanliness regime, ensuring that no infection threatens the health of her litter and no bad odour betrays the site of the earth. For the same reason, both parents take care to ensure that no telltale scats or food remains are left around the entrance.

This is an exhausting time for the dog Fox, who must continue to feed the ravenous vixen. It is at this time of year – March/April in the UK – that householders often report the theft of pet Rabbits and Guinea Pigs, and owners of such small pets should ensure that their animals are properly secured.

**Below:** The dog Fox continues to provision the nursing vixen with food.

# Growing up

Fox cubs grow up quickly. In their first month, confined to the den, they gain some 15–20g (½–¾oz) a day and by six weeks weigh just over 1kg (2¼lb). By four and a half months they have reached adult body size, but at around 3.5kg (7¾lb) are still only about half their parents' weight. Male cubs grow faster than females, tending to overtake their sisters at about four weeks.

After 4–5 weeks the cubs make their first venture above ground – typically, in the UK, on a warm day in April. They play exuberantly, leaping, chasing, biting and eventually slumping down to sleep in the open air. Play is an essential part of the animals' upbringing, allowing them to hone skills, develop muscles and refine instincts that will boost their survival as adults. It also allows the siblings to establish early pecking orders.

Once the cubs are up and about above ground, the vixen starts making longer trips to hunt for herself, often leaving the dog Fox lying close to the earth to keep a watch on her litter. At this time the cubs are vulnerable to predators – including cats, which in urban areas are drawn to the noises issuing from the earth – and the adults defend their young fiercely. The commitment of the male to this cause varies considerably: some males are doting dads and helpful husbands; others seem rather less interested and continue to pursue dalliances elsewhere. In general, the poorer the habitat and the leaner its resources, the stronger the dog Fox's commitment.

**Below:** Youngsters will play with solid food provided by their parents before they are weaned.

**Above:** Once out of the den, cubs are caught between the urge to stick with their mother and to explore.

Cubs are weaned at around 6–7 weeks. By this time they have a full set of milk teeth and can bolt down the solid food that the vixen provides – usually in the form of regurgitated meat. The vixen may already have been bringing solid foods for a few weeks, allowing the cubs to play with it and suck the juices, and she continues lactating for up to 12 weeks, but it is at this point – typically early May in the UK – that the all-important transition is made. To encourage the process the vixen keeps her distance during the day, and both parents continue to bring back small items of food. It is at this time, when there are small Foxes at the den entrance and no adult is in sight, that observers often believe the litter has been abandoned.

By 8–9 weeks – early June in the UK – the litter has left the earth and moved to a new 'play area', where the vixen leaves it when she sets out hunting. Left alone, the cubs show insatiable curiosity in the objects around them, taking small exploratory forays and making their own hunting attempts on beetles and other small creatures. This is the best time of year in which to watch Fox cubs. Adults may also bring them back toys, which in urban areas may include plastic toys, dog chews and even shoes – anything appealing and whiffy that they can filch from a patio.

By 10 weeks – mid-July in the UK – the cubs have become less visible and more wary. As the adults bring back less food, the cubs are obliged to accompany them on their hunting trips. By August, at about four months,

**Above:** Juvenile dominance hierarchies develop at an early age.

**Left:** Play-fighting between growing cubs helps them establish skills and status for later life.

the cubs are looking more like their parents and starting to keep adult hours – heading out between dusk and dawn. They still have a lot to learn, however, and rely upon their parents to sound the alarm with a sharp warning bark should any threat appear.

The cubs' first efforts at hunting are rather inept. Although you might see them practising the mousing pounce with great enthusiasm, it takes a few more weeks before they get results. Meanwhile they slake their appetites on easier fare, including earthworms – which are especially important in wet conditions, when they are easiest to capture – and blackberries. The latter come into fruit just as the cubs are learning to go solo, and provide a staple food during August and September, turning their droppings purple.

By the end of September the cubs are pretty much fully grown and hard to distinguish from their parents. They become more vocal at this time, caught between the twin imperatives of dispersal from the family home and trying to maintain contact with their parents and siblings, who are becoming increasingly hostile towards them. This is a testing time for the youngsters. They have not yet perfected all their survival skills, but with a hard winter ahead – their first time alone – they need to get into condition fast. Some 60 per cent of Red Foxes do not survive their first year. Although they reach sexual maturity in nine or 10 months (in around January), many do not reach breeding status or condition for another whole year.

**Above:** Anything that moves – including this duck – is fascinating to a growing Fox cub.

# Getting Along

Wherever they find themselves, mountainside or metropolis, Foxes need a place to call their own: somewhere that provides the basic survival prerequisites of food, water and shelter. The size and character of this home vary enormously, depending upon the habitat, its resources and the local competition. Wherever a Fox winds up, however, it can be sure that it will not be alone. The complexity of ways in which Foxes share their space – their social systems and communications strategies – are among the most fascinating things about this ever-surprising animal.

## Home on the range

A Fox needs to be sure that the area in which it lives can provide for its needs and is secure from intruders. The total area in which it operates, defined by the furthest extent of its normal daily wanderings, is known as its home range. This must provide it with enough food and sufficient hideaways in which to find secure dens and retreats. The size of a Fox's home range can vary from just 4ha (10 acres), in an urban centre chock-a-block with food and other Foxes, to more than 5,000ha (12,355 acres) in the middle of a desert, where food is hard to find. Home ranges in Europe vary on average from 40 to 1,300ha (99–3,212 acres). The largest home range known for Foxes in the UK is 4,500ha (11,120 acres), in the Scottish Highlands.

A home range is not necessarily the same thing as a territory. The latter may be a smaller area within the home range in which a Fox – or pair of Foxes, or small family group – locates its earth, and within which intruders from outside the social group are not tolerated. Linear landscape features such as hedges, lines of trees and city streets all make convenient boundary lines. During the spring and summer these are clearly defined and defended. In autumn and winter, however, once mature

**Opposite:** Territorial disputes may sometimes erupt into violence.

**Below:** Red Foxes like plenty of space.

**Above:** A pile of tyres makes a convenient territorial boundary marker.

cubs start to disperse and dog Foxes are off chasing females, boundaries become more porous.

Territories tend to be smallest in places where food is relatively plentiful, such as cities, and are more clearly defined and jealously guarded. Where food is exceptionally abundant, however, things can become more laissez-faire: the BBC *Springwatch* series offered a good example of this when documenting the fortunes of several Fox groups all sharing the rich pickings of Pitsea Landfill site in Essex.

Territorial borders are also constantly shifting. Studies in Oxford found that the city was divided into a honeycomb of Fox territories, but that as changes to the habitat – caused by, say, a new road or building development – caused one Fox to cede part of its territory, this was immediately occupied by a neighbour, which in turn lost part of its own territory to another Fox. Thus no one Fox ever ended up with significantly more land than its neighbour. When Foxes disappear from an area in numbers – as happened in Bristol during the mid-1990s, when a severe outbreak of sarcoptic mange (see page 89) killed some 95 per cent of resident Foxes – those that remain expand their territories to take up the slack. No space is ever left vacant.

## Keep out!

The ownership of a territory needs constant reinforcement. Foxes patrol their patch daily – not necessarily all of it every day but covering most of it, on average, every two days. They mark out its boundaries using smell: depositing scats or squirts of urine at strategic and conspicuous places such

**Below:** An interloper displays submissive body language while retreating from a territorial skirmish.

**Left:** A Fox marks its territory daily.

as rocks, tufts of grass and garden ornaments, and anointing them with a pungent secretion from their anal glands. When scent marking with urine, a male Fox raises one hind leg and sprays his urine forwards, whereas a female squats down and sprays on the ground between her hind legs. Either way, these olfactory signposts get the message across just as forcefully as any gangland graffiti. 'Keep out!' they tell would-be trespassers. This greatly reduces the chance of rivals meeting and perhaps coming to blows.

Scent marks do more than simply reveal that a territory is taken. Each scat or squirt comprises a cocktail of chemicals that speak volumes about the resident animal's sex, health, relatedness and social status. The wind and rain whittle away at the scent, so Foxes make sure to top it up at regular intervals. Any interloper hoping to sneak by undetected usually plots a course through the territory using the oldest, stalest scent marks.

Inside their territory Foxes move around confidently, following the same well-trodden paths between earth and feeding ground. You can sometimes spot these 'Fox highways' by the line of flattened vegetation. They often follow ready-made trails, such as vehicle ruts or the bolder tracks of Badgers. In a neighbour's territory, however, a Fox will move cautiously, sneaking from one patch of cover to another and not scent marking for fear of giving itself away.

On occasion, notably during the courtship season, Foxes wander right out of their home range. Studies in

**Below:** The quills in this strategically placed Fox dropping reveals that birds have been on the menu.

**Above:** Elevated points, such as this hay bale, are good places on which to leave a pungent message.

the US found that some dog Foxes travelled up to eight times as far during this period, while the home range of some Foxes in Bristol doubled in size. Foxes may also sometimes exit their home range entirely to visit an irresistible food bonanza beyond their borders – such as, in Japan, a pile-up of salmon carcasses along a river.

Territory is not for everyone. Some 15 per cent of Foxes – mostly males – have no home range at all but are constantly on the move. These individuals are known as itinerants. The greatest nomad on record in the UK was a dog Fox from Brighton, Sussex, that travelled east to Rye in less than a month over the winter of 2013/14, covering some 315km (196 miles) in the process. But even this impressive journey falls short of the record of 394km (245 miles) recorded in Wisconsin, USA.

**Below:** A dominant Fox is confident within its territory.

# Living together

The wily Fox of popular culture is a loner – a solitary thief in the night who keeps himself to himself. However, studies in the UK during the 1970s by celebrated Fox experts Stephen Harris (in London) and David Macdonald (in Oxford) revealed a rather more complicated picture.

It is true that Foxes as individuals are mostly solitary. Adults operate largely alone, except when with their young, and even an established pair spends most of its time apart. Certainly the idea sometimes put about by tabloids that urban Foxes hunt in packs is a myth. While other canids, such as Grey Wolves and African Wild Dogs, have evolved communal hunting as a means of bringing down larger prey (a Wolf could not tackle a Caribou alone), Foxes are primarily hunters of small rodents. These bite-sized morsels are meals for one. Indeed, there is nothing worse in Fox society than approaching another individual while it is eating. Being solitary, however, does not mean having no social life, and Foxes have evolved their own modest take on group living.

A Fox group is a nuclear family made up of only blood relatives. It consists typically of a dominant dog Fox – the 'top dog' – and a dominant vixen, together with around four or five (occasionally up to 10) subordinate vixens from previous litters. It is very rare that the dominant dog Fox will tolerate other males in the group. His own male

**Above:** An adult Fox spends much of its life alone.

**Below:** A vixen with two of her cubs. Female offspring may remain with her for a year or more.

**Above:** A dominant vixen asserts its authority over a younger female.

**Below:** Young adult Foxes tussling for position.

cubs are chased out as soon as they are ready to fend for themselves, and disperse to territories elsewhere. This guards against the dangers of inbreeding.

Every Fox group enforces a strict pecking order. The dominance hierarchies are established from as early as 12 days, when the cubs first open their eyes and siblings start pushing and shoving for access to the vixen's teats. These battles continue during play once they have left the earth. As soon as a structure is established – with an alpha 'top dog' heading the hierarchy and a runt at the bottom – any further serious violence is largely avoided.

The dominant pair is responsible for maintaining and defending the territory. It does the bulk of the scent marking and sometimes works together to see off an intruder. Lower-ranking vixens get on amicably for most of the time, grooming one another to reinforce bonds, but relations are in constant flux and can quickly deteriorate. Tensions often come to a head during autumn, when cubs are dispersing and the dominant pair becomes increasingly intolerant of subordinates.

# Cub killers

It is an unpalatable reality that Foxes occasionally kill cubs – either their own or, more commonly, those of another Fox. This behaviour, known as infanticide, is known in many mammals, from Lions to Rabbits, Chimpanzees and of course humans. Most instances in Foxes have been recorded on fur farms, especially among Silver Foxes. Only a handful of verifiable cases are known from the wild, but because Fox cubs are out of sight in the dens during their first few vulnerable weeks, it is possible that the habit is more common than we think. In a few rare cases the killer may eat – or at least partially consume – an unfortunate youngster.

The reasons for infanticide are complex and the jury is out on its cause among Foxes. Various theories have been advanced. One is that the killing is purely predatory – in other words, for food. Another is that it aims to remove potential competitors for resources. A third is to do with sexual selection – males killing the young sired by other males to prevent them from raising kits that are not their own. Other explanations suggest that infanticide has no evolutionary benefit and is just an occasional aberration caused by a surfeit of stress or aggression – in other words, the poor cub was in the wrong place at the wrong time.

In instances of infanticide in the wild, it is very difficult to draw conclusions without knowing the relationship between the adult killer and the infant victim – which is usually the case. It seems, however, that the sexual selection hypothesis – used to explain infanticide in Lions, whereby the killing of cubs by a new male entering the pride brings females back into oestrus and thus able to mate with him – is less likely to apply to Foxes, where the female has a fixed breeding season. In many cases it seems that competition may be the main driver, with predation a side effect if the killer happens to be seriously hungry.

**Above:** A female American Red Fox carries away a dead cub from the earth, watched by another cub.

# Mother's little helpers

Subordinate female Foxes get plenty out of living in a group, notably the guaranteed food and security that comes from having an established territory – but they also make a major sacrifice: they never get to breed. The dominant female effectively denies their reproductive opportunities by harassing them constantly, controlling their access to food and generally keeping them in their place. The stress of this unremitting subordination keeps their reproductive hormones suppressed. They can only take so much, however, and at some point – when the dominant vixen takes her bullying too far – they may get fed up and strike out alone.

The dominant pair, meanwhile, benefits from having family around. The subordinates not only help to defend the territory, but also provide free childcare services. The younger, lower-ranking females – all grown offspring from previous litters – act as 'helpers' to the new litter, playing with the cubs, grooming them and providing them with food. They have even been known to lactate and suckle their charges. This behaviour kicks in at about four weeks, once the cubs have left the den; until

**Below:** A vixen greets a rather wet young cub.

then the dominant female is fiercely protective of her new litter and will attack any other adult that comes too close. Such behaviour is known as alloparenting. It is relatively unusual among mammals, because it generally involves the helpers sacrificing their own reproductive opportunities. Other mammals among which it is known range from Warthogs to jackals.

**Above:** A Fox cub begs for food from its parent.

**Below:** A vixen with her three-month-old cub.

# Spreading the word

**Above:** A pair of Foxes greet each other with chattering calls.

Communication is vital in Fox society, both within the social group and outside it. The message may be a hostile one, especially between territorial rivals, but this is better than a fight, which can be costly in terms of both injury and energy – and is a waste of time that could have been spent feeding. Thus Foxes use a complex language of both sound and gesture to get the message across. Watch several individuals together and you will not only hear an impressive range of vocalisations, but also witness the ever-changing ballet of choreographed posture, movement and expression by which they manage their relationships with one another.

## Body language

Foxes are highly articulate with their body language, and most dog owners will find at least some of it reasonably easy to interpret. For instance, standing tall, with tail erect, ears pricked and whiskers raised signals dominance. By contrast, crouching low, with head lowered to the ground, ears flattened against the top of the head, mouth agape and tail curled under the body, is an expression of submission. There are many variations on these, according to circumstance. A vixen rolls on her back to expose her vulnerable genitals and stomach as a gesture of submission to a dominant male. She may also squirm at his feet, lick his muzzle and thrash her tail. Arching the tail in an upside-down 'U' and flattening the ears signals an invitation to play, but crouching with ears flattened and hackles (hair on the shoulders) raised signals an intention to attack.

Facial expression also helps. A direct stare is an assertion of dominance. Eyes narrowed and turned away, by contrast, signal submission. An open mouth is not necessarily a sign of aggression; if opened just a little, with lips drawn back horizontally, it is a play face. Foxes do not snarl – despite the word 'snarling' being a prerequisite of every alarmist tabloid headline. Images that appear to depict snarling are often either of badly stuffed animals, or snapped while a Fox was in mid-mouthful.

These ritualised expressions generally serve to avert any serious conflict. When two individuals reach a stalemate, however, a separate ritual unfolds. The two animals stand side by side in a competitive assessment of size and confidence, their backs arched, tails curled to one side and heads turned away. At this point one

**Below:** Foxes use a variety of body language to assert status, and convey dominance and submission.

combatant usually recognises its inferiority and slinks away. If neither backs down, however, things may escalate to pushing and shoving in a flank-to-flank test of strength. If this still does not work, an all-out fight may erupt, the two combatants trading savage bites to the head and rump as they aim to seize hold of the opponent's neck. Such serious fights are rare. When they do occur it is invariably between members of the same sex; there is seldom any violence between male and female.

**Above:** A 'foxtrot', in which two combatants rear on their hind legs and wrestle, may look fierce but seldom gets too serious.

Vixens and younger Foxes sometimes fight by dancing a 'foxtrot'. Two combatants rear up on their hind legs, place their forepaws on each other's shoulders and, mouths agape, attempt to topple one another to the ground. The loser is the first to topple, but sometimes both parties end up on the ground together, thrashing about and screaming. This may look violent but serious blood is seldom spilt. Indeed, apparent violence in Foxes is often not what it seems. What may appear to be a vicious conflict may actually be a bout of play-fighting. In younger Foxes – as in domestic dogs – what starts as a fight may quickly end in play.

## The calls of the wild

Foxes can be very noisy. Scientists have identified at least 20 distinct vocalisations, including eight used only by cubs. Each serves a different purpose and the repertoire broadens as a Fox matures. Individual Foxes can recognise one another – and identify strangers – from these calls.

The best-known – and certainly the loudest – call is the scream. This high-pitched, hoarse and rather tortured sound is most often heard during the breeding season and is sometimes mistaken for the cry of a woman in distress, even prompting worried householders to call 999. It is often described as a vixen's mating call, but is used by both sexes and serves generally as a contact call to establish the whereabouts of a partner. Another common call is a softer, more rhythmic bark – a kind of *ow-wow-wow-wow*, sometimes mistaken for an owl's hooting – which Foxes use to mark territory. They also employ a yapping, more staccato bark for long-distance contact between group members.

Most other Fox vocalisations are quieter and used for communicating in closer proximity. A common call, also quite bird-like, is a high-pitched, guttural chattering, interspersed with occasional yelps and howls, known as gekkering. Adults use this in aggressive encounters, as do young kits when play-fighting. Submissive Foxes sometimes emit piercing whines when greeting their superiors; the whines can turn into louder shrieks if they are feeling especially subordinate.

**Above:** The fox's high-pitched 'scream', often attributed to vixens, is a contact call used by both sexes.

A vixen, meanwhile, has a whole separate language for her cubs. A rumbling, cough-like bark alerts youngsters to danger, sending them scurrying for the safety of the earth. A low growl, sometimes described as a mew, churr or purr, summons them to her. Studies have shown that a vixen can call individual cubs, who respond accordingly, suggesting that each cub effectively has its own name. Down in the den, a vixen communicates with low huffing and coughing, and brief clucks, like a kind of maternal gekkering. Newborn kits start off with a whelping noise, which by three weeks has become a more rhythmic, insistent yelping. This escalates to a kind of high-pitched warbling, as they demand more and more attention from their mother.

**Below:** Young kits play-fighting can be very noisy.

# Dangers and Disease

The Red Fox may seem the very paragon of success: a resourceful, live-anywhere opportunist; the world's most widely distributed mammal, and so on. However, this does not mean that it has things easy. On the contrary, a Fox's life is a short and hazardous one that generally ends brutally. As if our own persecution of the animal is not enough, it seems it must cope with everything from predators and parasites to pollution and road traffic. Its success, therefore, lies not in a lack of adversity but more in an uncanny ability to beat the odds.

Captive Red Foxes may jog on to a ripe old age, with the record belonging to one individual in Utah, USA, which reached 23 years and 7 months. Longevity records in the wild are rather harder to verify. They include records of a Fox in Hokkaido, Japan, which died at 15, of one in Switzerland that reached 13 and of a radio-collared individual in Holland that was killed aged 12 by a policeman.

None of these exceptional statistics, however, should disguise the fact that the lifespan of most adult wild Foxes is from two to six years. In the UK only some 5 per cent of Foxes live past their fourth birthday, with a slightly higher life expectancy among those in rural areas. One study of urban Foxes in Oxford found that 63 per cent die during their first year and only 12 per cent reach their second birthday, giving an average life expectancy for that city of just 19 months.

Cubs have it the hardest. In most regions more than half die within their first year. Many do not make it past their first few weeks, with roughly one-fifth of cubs dying underground, usually the victims of sibling squabbles over milk and food. Of 1,000 Fox cubs studied in one Bristol survey, 15 per cent died at four weeks, 65 per cent survived to sub-adulthood and only 39 per cent made it to adulthood.

**Opposite:** The Golden Eagle is the only UK predator that regularly – albeit rarely – includes the Red Fox on its menu.

**Below:** Fox cubs are at their most vulnerable during their early weeks.

# The perils of people

**Above:** Fox killed in a gamekeeper's trap.

Many things can kill a Fox. Threats vary in their severity with such factors as age, social status, habitat, location and population density, but the most significant dangers across much of the Fox's range come in human form – either via direct hunting and persecution, or from the numerous other more insidious ways in which we have made its world unsafe.

Today in the UK there are laws that govern *how* you may kill a Fox: it is no longer legal, for example, to poison or gas Foxes. However, killing Foxes in general – provided you have obtained the appropriate licence – remains perfectly legal. An estimated 80,000 Foxes are shot in Britain every year, about half by gamekeepers. Many others are killed in snares, by terriers or at their earths.

Deadlier than any rifle or snare is the motorcar. Traffic accidents kill an estimated 100,000 Foxes every year in the UK alone. One survey in Oxford found that this was much the biggest single cause of mortality in that city, with around 60 per cent of Foxes meeting their end on the city's roads. Statistics reveal that cubs aged 3–7 months are most at risk from vehicles, with traffic casualties reaching a peak during autumn, when these youngsters are dispersing from their natal areas. When you consider that the average annual UK winter population of Foxes – after the later spring peak of cubs – is some 250,000 individuals, you need not be a mathematician to get some idea of the scale of the slaughter we inflict on the species.

**Below:** Crossing a road is a dangerous business.

Then there are the numerous other hazards of the human environment. Our poisoned landscape drips with rodenticides and crop-spray insecticides, which find their way into Foxes along the food chain to cause blindness and other damage. Pollutants such as heavy metals and PCBs accumulate in body tissues, passing through a vixen's milk and compromising the health of her cubs. Ironically, urban Foxes tend to suffer fewer harmful effects from pollution than their rural cousins, as the human foodstuffs they scavenge are much lower in pollutants than the substances that we spray over our countryside. However, these city-dwellers must contend with an assault course of physical hazards in suburban back gardens, including netting, fences, swimming pools and ponds, with many cubs becoming tangled, choked or drowned.

**Above:** As many as half the adult Foxes in some areas of the UK end up as traffic casualties.

**Below:** A countryside of monocultural crops and pesticide pollution poses many threats to Foxes.

**Above:** A dead Red Fox trapped in a wire snare.

**Below:** A line of dead Foxes strung on a fence does not act as a deterrent to other Foxes, no matter what the farmer may imagine.

# Wild enemies

Plenty of other animals can and do kill Foxes. The biggest threat comes from rival predators. These include – depending upon where you are in the world – bears, Wolves, lynxes, Leopards, Cougars, Coyotes, Dingos and Wolverines. In most cases, however, the killing is not due to active predation but rather to the elimination of competition. In other words, these larger animals are not hunting the Fox in order to eat it, but aiming to remove a fellow predator that is competing for the same food resources. They generally leave the carcass of their rival alone.

Coyotes and Foxes have a particularly troubled relationship. In the rural USA, scientists from California have calculated that Coyotes are responsible for 38 per cent of Fox deaths – and there is no doubt that a percentage of these are consumed as food. In Yellowstone National Park the recent reintroduction of Timber Wolves has led to a fall in Coyote numbers and, conversely, a rise in the Red Fox population. Relations are not inevitably hostile: Coyotes and Red Foxes have also been observed feeding together from the same carcass – as have Striped Hyenas and Foxes in the Middle East and southern Asia, although a Fox will stay well away from the bone-crunching jaws of hyenas. Stray dogs play a similar role to Coyotes in some regions, and the high population of feral dogs in many parts of the developing world partly explains why Foxes have not adapted so readily to urban life there.

**Above:** Coyotes pose a significant risk to Red Foxes in parts of the USA.

**Above:** A Red Fox in Kamchatka, Russia, watches warily as a Wolverine ambles by.

**Above:** The Northern Lynx will make short work of any Fox it can catch.

**Above:** Two Golden Eagles squabble over a Red Fox carcass.

Among the very few carnivores that actively hunt Red Foxes for food, one of the deadliest is the Golden Eagle. This powerful raptor seldom takes prey of more than 4kg (9lb) in weight, since it cannot fly carrying a heavier load, so its attacks are mostly confined to subadult Foxes and cubs. However, it is perfectly capable of killing an adult, and may feed *in situ* on prey that is too large to carry. An eagle captures a Fox in a low-level swoop, striking with its talons at the animal's neck and clinging on until it collapses.

The Fox is not a major part of the Golden Eagle's diet, comprising no more than 2–4 per cent of its prey in most European regions. This is unsurprising: predators such as Foxes are hard and dangerous to capture, and occur in much lower densities than other prey – especially in the remote regions favoured by the eagle. Nonetheless, researchers have found Fox remains in more than 50 per cent of Golden Eagle eyries in some areas. The raptor's predatory prowess has been harnessed by the Kazakh people of the Altai Mountains in western Mongolia, who for more than 200 years have trained Golden Eagles to hunt Foxes on the open steppes (see opposite).

# Eagle hunters of Mongolia

Hunting foxes with Golden Eagles is a traditional art of the Eurasian steppes, practised today by the Kazakh people of the Bayan-Olgii Province of Mongolia. The practice, known as *berkutchy* in Kazakh, has ancient roots among the nomadic Khitan dynasty, who conquered part of northern China in 936 BC–AD 45, and indeed falconry itself is thought to have originated in central Asia more than 5,000 years ago. Today there are some 250 eagle hunters who pursue this unique form of horseback hunting. For these individuals, who inherit the tradition from their fathers, it involves a lifetime's dedication. Their relationship with the Eagle is all-consuming, the hunter going without sleep for long periods during the training and keeping the bird unsighted under a hood until its dependence upon the human becomes complete. Bonds between hunters

and their Golden Eagles, which are very long-lived species, may last for more than 20 years. The preferred targets are Red Foxes and the smaller Corsac Foxes (see page 28).

Most hunting takes place during winter, when snow makes the foxes easiest to see. The hunter takes his bird out by horseback to a high vantage point and releases it from the wrist when a fox is sighted. The Eagle swoops low and strikes the fox on the back of the neck, pulling it to the ground instantly or 'riding' it until the animal collapses. Sometimes the Eagle kills the fox outright; at other times the hunter arrives on horseback to dispatch it. The Eagle is then lured off its prey with a piece of meat as a reward. An annual Golden Eagle Festival is held in the first weekend of October: up to 50 hunters compete for a variety of prizes awarded for speed, agility and accuracy.

**Above and below:** The Kazakh eagle hunters are famed for their horsemanship, and develop an extraordinarily close bond with their birds.

**Above:** A German spaniel stands guard over a hunter's quarry.

**Below:** A Fox would do well not to come between a deer and its fawn.

Apart from the Golden Eagle, other birds of prey known to hunt Foxes include Bonelli's Eagle, Steller's Sea Eagle and the European Eagle Owl – in all cases they generally take cubs. Young Foxes in the den are also vulnerable to other hunters, including Mink, skunks, Stoats and Badgers (both American and European). Domestic dogs will kill and eat Fox cubs – as will some domestic cats, given the opportunity. In the case of the latter, this suggests that the aggressive behaviour of Foxes towards cats during the breeding season is born more of defence than attack. Even deer have been known to kill Foxes in defence of their fawns, trapping the predator in a thicket and using their horns and hooves to gore and trample it to death.

Finally, Fox may turn upon Fox. Infanticide is an unpalatable but widespread reality in Fox society (see page 71), and adults occasionally – though very rarely – inflict fatal injuries upon one another during territorial battles. Such deaths are not acts of predation, although during hard times one Fox is not above scavenging the carcass of another. Life is hard enough without letting good food go to waste.

**Above:** A Red Fox risks the wrath of a Black Vulture while scavenging from a carcass in the mountains of central Spain.

**Below:** Fox-on-fox violence can sometimes prove fatal for one party.

# Injury and illness

**Above:** An injured Fox receives treatment at a rescue centre.

Foxes, like all wild animals, get hurt. One UK study found that some 30 per cent of male Foxes and 35 per cent of females had at least one healed bone fracture. Many also suffer bullet wounds and lose feet or even limbs to snares and fences. The capacity of Foxes to survive such damage is remarkable. Healthy, functioning individuals have been observed with, among other horrors, a bullet wound passing straight through the skull, an arrow protruding from the chest and a snare embedded around the waist, around which the skin had grown back. A one-eyed Fox raised two litters of cubs before she developed a cataract in the remaining eye and was hit by a car.

Of more concern to the health of many Foxes is the parasite load that most carry. These hangers-on include a variety of intestinal worms, flukes, ticks, fleas and mites, and it is a parasitic mite that causes the common, debilitating and invariably fatal condition known as sarcoptic mange (see box, page 89). One of the oldest myths about Foxes – dating to ancient Greece – concerns their alleged ability to shed fleas by entering the water backwards, holding in their mouth a piece of bark or sheaf of grass. The fleas, naturally, hop up the Fox's body to escape the water, ending up on the piece of bark, whereupon the now flea-free Fox releases its tormentors to drift away in the stream.

**Below:** A Fox swimming to rid itself of fleas – if the ancient myth is to be believed.

# Mange

Many householders have encountered the distressing sight of a Fox with mange. The animal has generally lost lots of fur, especially from the rump and tail, leaving the exposed skin bald, encrusted and raw, and its eyes may appear half-closed with conjunctivitis. This condition, properly known as sarcoptic – or canine – mange, is as miserable as it looks. An infected Fox loses condition and weight rapidly, suffers organ damage and is likely to die within 4–6 months. The wider consequences can be catastrophic: an outbreak of mange in Bristol during the 1990s killed 95 per cent of the city's Foxes within two years. Another outbreak in Australia during the 1940s and '50s led to an 80 per cent crash in the population.

Contrary to popular belief, mange is not confined to urban Foxes; indeed, the Bristol outbreak was caused by an infected rural Fox that arrived from outside the city. The culprit is a tiny parasitic mite, *Sarcoptes scabiei*, which burrows into the Fox's skin in great numbers and may live there for a month. The deposit of tissue fluid and debris left at the surface causes the

Fox intense irritation. It scratches away, losing hair, lacerating the skin and opening itself up to bacterial infections. The mites are spread easily between animals, either directly (the sniffing of hindquarters in greeting explains why the condition is often concentrated at the rear end of a Fox), or via a shared food source or bedding area.

Mange is not confined to Foxes. It can spread to other animals, including some domestic dogs – although experiments in the US have shown that many animals, including Rabbits, Raccoons, skunks, opossums and cats, were not susceptible. The condition is treatable. Vets and animal-welfare charities use an anti-parasitic medication called ivermectin that can be administered orally through food, although this does pose a risk to some breeds of dog. There are also a number of homeopathic alternatives. The scientific jury is still out on their long-term effectiveness, but they do not affect domestic livestock and some have a proven success rate among Foxes. The National Fox Welfare Society (www.nfws.org.uk) distributes thousands of free packs of one homeopathic treatment every year.

**Below:** The first signs of mange in a Fox often appear around the rump and tail.

**Below:** In severe cases, a Fox may lose virtually all its fur.

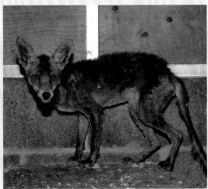

# A deadly bite

Most diseases that affect Foxes are of little concern to humans. Foxes may contract *Toxocara canis* (Dog Roundworm), bovine TB and Weil's disease, all of which can have a serious impact on people and/or livestock, but they are not thought to be a significant vector in the spread of any of these. The one serious exception, however, is rabies. Our fear of this disease, which can be fatal in both animals and humans, has had a major impact on our relations with Foxes over the last century.

Rabies is a viral disease of the central nervous system transmitted via the saliva of infected animals. In livestock and domestic animals it is known as 'classic' rabies; in wild animals, 'sylvatic' rabies. Victims experience – among other symptoms – violent convulsions, a fear of water (hydrophobia) that is prompted by these convulsions, aggression, disorientation and paralysis. The virus may incubate for 2–8 weeks before symptoms appear. Treatment for humans – a vaccination – is effective if administered during this period. However, once the symptoms appear the disease invariably proves fatal. Other animals succumb more quickly, usually dying within four days. An infected animal may become extremely aggressive – the rabies virus *Lyssavirus* gets its name from the Greek *lyssa*, for 'frenzy' – but not always: a lick from an apparently docile infected animal is enough to spread the virus, should it enter an open wound.

In much of the world the Red Fox is a key carrier of rabies – spreading it to domestic dogs – and for centuries it has been synonymous with the disease. The first mention of rabies in the UK dates back to 1026. In 1793 a ban was imposed on importing dogs into the country, and in 1819 the Duke of Richmond allegedly died after his pet Fox licked his face (the saliva reputedly entered his shaving cuts). In 1897 a General Rabies Order required the muzzling of all dogs in public. In 1922 the disease was declared absent from the UK. Since then there have been a handful of outbreaks,

**Above:** Preparing the rabies vaccine for local dogs in Scoresbyssund, Northeast Greenland.

all quickly contained. The most recent death was that of a young Scottish scientist who, in 2002, contracted rabies from one of the bats he was studying. A compulsory quarantine introduced in 1922 for any dogs moving in and out of the UK has now been replaced with a pet passport system, which allows the movement of animals that have proof of vaccination.

The control of rabies in Foxes is seen as one of the great triumphs of 20th-century wildlife management. The initial strategy was culling. However, this was ineffective in many areas, because the Fox population was quickly able to recover and reoccupy any vacant territories. Vaccination has since proved much more effective. This technique was first trialled in the Swiss Alps in 1978, by means of smuggling the vaccine into bait. It was found that if more than 60 per cent of the bait was taken, the Fox population built up an effective wall of immunity. This approach has helped to eradicate rabies from much of western Europe, including France and Germany. Nonetheless, it remains widespread in eastern Europe and scientists stress that there is no room for complacency. In North America rabies is associated more with Raccoons than Foxes, but similar immunisation programmes have been undertaken in both animals.

**Above:** A rabies warning sign in Germany.

**Below:** An Ethiopian Wolf – another wild canid susceptible to rabies – is released by researchers in Bale Mountains National Park, Ethiopia, having been trapped and tested for rabies antibodies.

# City Slickers

'So, are you pro or anti?' This was the instant response when
I told a friend that I was writing about urban Foxes, and it
encapsulates the difficulty of the subject. The Foxes that have
taken up residence in our towns and cities are not simply a
natural history phenomenon; they are officially an 'issue' – and
one about which everyone is expected to hold an opinion. One
thing is for sure, however: the urban Fox is here to stay. If we
are to learn to live with it, we would do well to find out a little
more about it. In recent years the media has fanned the flames
of the debate, polarising opinions still further. To some, urban
Foxes are a disease-spreading, litter-scattering, pet-murdering
menace. To others they provide company and entertainment
and bring a welcome breath of the wild to our sterile city
streets. Amid the controversy, fact can be hard to separate
from fiction.

# Heading into town

Urban Foxes are not as recent a phenomenon as is often
thought. Foxes have been roaming city streets in the UK
since at least the 1930s, when sightings in the leafier
parts of London – including Hampstead Heath and
Richmond Park – first began to appear in the records.
After the Second World War the trend continued, and by
the 1970s Bristol and Oxford were among several other
cities to have established thriving populations. Today most
large UK towns and cities are home to Foxes. Precise
numbers are hard to calculate – and populations fluctuate
significantly, most notably due to outbreaks of mange,
which caused Bristol's Fox population to fall by some
95 per cent during the mid-1990s – but recent estimates
suggest an overall figure of 35,000–45,000: some 14 per
cent of the overall UK population.

**Opposite:** City streets offer a
highly productive habitat to the
Red Fox.

**Below:** Pavements serve as well
for Foxes as for people.

**Above:** Foxes find numerous niches in urban environments: (from left to right) climbing on a roof; crossing a park in broad daylight; investigating a fallen bin.

Numerous ideas have been put forward to explain this trend. It was once thought that the departure of gamekeepers from country estates during the Second World War allowed Fox numbers to rise so high that the animals were forced into town in search of more space. Another theory held that the advent of myxomatosis in the 1950s caused Fox populations to soar, fuelled by all those dead Rabbits, and then – when no live Rabbits replaced them – to leave the countryside in search of food in the city. Some have suggested that the Fox-hunting ban of 2005 allowed Fox numbers to rise and, again, spread to the city – although the data has not shown any rise in the UK Fox population since the ban came into effect.

A better way to explain this might be to consider that perhaps the city came to the Fox, rather than vice versa.

**Below:** Railway lines have played an important role in enabling Foxes to colonise suburbia.

A boom in housing development during the 1920s–1940s, and especially the trend towards owner-occupied housing away from industrial centres – in other words, the mushrooming of suburbia – saw urban areas expanding into what had once been the Fox's rural kingdom. Foxes, with their great ecological resilience, adapted to their new home. They could now supplement their diet of rodents and 'natural' prey by scavenging from the huge amount of waste left around by people. They could also find secure den sites in wasteground, along railway embankments and under sheds – back gardens becoming ever more appealing due to the recent garden-makeover fashion for decking and other structures with cavities beneath them. The threats were minimal: traffic could take a heavy toll, but city people were generally far less concerned than their country cousins about having a Fox in their garden – for most, it was of little relevance to their lives; for some, it was a source of pride.

**Above:** A Fox cub peeks out from its earth beneath a garden shed.

## Around the world

Another misconception is that urban Foxes are a purely UK phenomenon. The last three decades have seen Foxes establish themselves in cities all across Europe, including Paris, Rome, Stockholm, Oslo, Berlin, Stuttgart, Zurich, Geneva and Copenhagen. Although in much of mainland Europe the trend started a little later than in the UK, numbers have since caught up fast. In Switzerland, for example, Foxes were first recorded in Zurich only in the 1980s. By 2004 they had colonised every major town

**Above:** A vixen suckles her litter in broad daylight under a block of flats.

across the country, and Zurich itself now has an estimated 10 Foxes per square kilometre.

Urban Foxes make their homes in numerous other cities worldwide – from Los Angeles, New York and Washington in the USA, to Toronto in Canada and Hokkaido in Japan. Australia has a particularly prolific population, with the animals being common in all southern and eastern cities. Indeed, Foxes were already well established in Melbourne by the 1930s – the same decade that saw them gain a foothold in UK cities – and today, by some estimates, that city is thought to have the highest density of urban Foxes anywhere in the world.

In parts of the world where rabies remains a problem, including eastern Europe, Foxes are more heavily persecuted and have not made the same inroads into urban areas. The packs of feral dogs that roam the urban areas of some developing regions, such as southern Asia and the Middle East, also make these areas less attractive to Foxes.

**Right:** Waiting for a break in the traffic.

# An unwelcome visitor?

Unfortunately, the arrival of the Red Fox in our towns and cities has not been met with universal celebration. Among the 'crimes' of which the animal stands accused are damaging our property, fouling our streets, spreading disease, keeping us awake at night and conducting a reign of terror over our pets and children. If the pejorative adjectives applied to rural Foxes are generally 'wily' and 'thieving', then the equivalents for their city cousins tend more towards 'filthy', 'savage' and 'starving'.

Our understanding of how Foxes live among us has not benefitted from inflammatory tabloid headlines and the pronouncements of politicians in search of populist causes. 'They may appear cuddly and romantic,' said Mayor of London Boris Johnson to *BBC News* in February 2013, 'But foxes are also a pest and a menace, particularly in our cities.' Amid the scare stories, facts can be hard to find, so it is worth looking more closely at the Fox's charge sheet.

Certainly Foxes can and do damage property. They dig beneath garden sheds, into lawns and under fences, tear though netting and plastic sheeting, and steal plastic toys, dog chews, footwear and other appealing items for their cubs. They also filch fruits, set off security lights, excavate buried pets and litter gardens with their pungent droppings. There are numerous ways to deter them (see box, page 100), but many householders are happy to put up with the odd unsightly excavation or chewed wellie for the pleasure of playing host to these captivating wild animals.

**Above:** A greenhouse offers a sun-warmed refuge for a daytime nap.

**Below:** Almost any discarded human object may be of interest to a Fox.

# Shoe fetish?

In June 2014, according to *BBC News*, Leeds householder Elaine Hewitt reported that dozens of single shoes – ranging from sandals to work boots – were turning up outside her house. When the rate escalated to a shoe a day, Elaine put up a shoe rack outside her home in the hope that neighbours would reclaim their missing footwear. They came from far and wide. The shoe thief, caught in the act one day, turned out to be a vixen with five cubs. Many similar stories have been reported from around the UK and as far afield as Germany. Often the footwear is not chewed or damaged, so clearly it is not mistaken for food. Foxes, it seems, have a kleptomaniac fascination with manmade objects – and not just shoes, but any bright-coloured, portable objects, such as balls, children's toys, ornaments or gardening gloves that might be left out around a garden. Plastic and leather are generally the preferred materials, but among the more bizarre thefts recorded are golf balls from a golf course – during a round – and 36 steel petanque balls taken from a garden in France, each weighing 800kg. Possible reasons for this

behaviour including Foxes honing their hunting and caching skills, or collecting objects for their cubs to help them do the same. Whatever the explanation, if you have foxes in your area, especially over the summer, you might do best not to leave your shoes – or anything else you value – outside overnight.

**Above:** A garden gnome warrants thorough investigation.

**Below:** Leaving a bin uncovered is asking for trouble.

As for the mess, yes, Foxes do raid bins. However, they are not as dependent upon human food waste as is often supposed, pursuing their natural predatory habits – catching rodents and other small prey – in towns just as they do in the countryside. What is more, the modern wheelie bin, with its height and closed lid, thwarts the efforts of most Foxes. Unprotected bin bags left torn apart, their contents strewn across the street, often represent the work of other culprits, including cats, stray dogs and, in seaside towns, gulls. Moreover, balanced against any charges of unhygienic waste spreading levelled at Foxes should be the pest-control service that they perform on our behalf – not only clearing up the waste itself, but also preying on the rats, mice, pigeons and other animals associated with it.

# Pets under attack?

The debate about urban Foxes becomes more emotive when it involves attacks. However, the evidence suggests that the danger Foxes pose to our household pets has been greatly exaggerated. Dogs, certainly, are at little risk. Beyond the demise of a two-year-old Chihuahua in Poole, there are no recent records of Foxes causing any serious harm to man's best friend. Indeed, being little more than cat-sized themselves, Foxes are very wary of most dogs and give them a wide berth. Dogs often dig out and kill Fox litters, and in areas where stray dogs are common Foxes tend to be rare.

Cats, you might think, are more vulnerable – and certainly cat owners' fear of Foxes runs deep. There are records of Foxes attacking and killing cats, generally involving kittens. However, these incidents are rare, and most evidence suggests that cats more than hold their own in the presence of Foxes. Indeed, where the two animals meet – which is usually after dark – the cat tends to be dominant, with the Fox not daring to risk injury. A big tom cat, after all, weighs about as much as a vixen and is equipped with lethal weaponry. Numerous amateur videos (for example on YouTube) testify to the respect that Foxes show cats. Furthermore cats – being predators themselves – are a genuine threat to Fox cubs, especially when the mother is not around. Incidents of Foxes acting aggressively towards cats in around March–April often reflect Foxes trying to protect their young in the earth.

In the event of a conflict it is easy to jump to the wrong conclusion. A bloodcurdling fight is heard during the night, a bloodied moggie limps through the cat-flap and a short while later a Fox is seen trotting across the lawn. Who knows what happened? The assailant might have been another cat, or a stray dog, and the Fox was just a spectator. Similarly, the remains of a cat found in a Fox's earth or during an analysis of its stomach contents do not prove that the Fox was a killer. Foxes are scavengers and will not turn down a good bit of road kill, feline or otherwise.

**Above:** A large cat is seldom intimidated by a Fox.

# Outfoxing the Foxes

If you prefer not to share your garden with Foxes, there are various measures you can take to deter them. Be warned, however, that these resourceful animals can get past all kinds of defences and deterrents, and some Foxes are more sensitive – or more persistent – than others. The trick is to make their visits less rewarding. The following guidelines might help.

•   **Pet protection** To keep your rabbits or guinea pigs out of a Fox's hungry jaws, house them securely at night. Make sure that the hutch or enclosure has a study frame and is made from steel mesh, not chicken wire or plastic mesh. It should have either a secure floor or wire buried into the ground to prevent Foxes from digging in from below. Fasten the door with bolts, not wire twist catches.

•   **Digging disincentives** Foxes may dig in your garden to find or bury food, or to excavate an earth. Avoid using fish and bone fertilisers. The smell of these fertilisers may also cause them to dig in the hope of unearthing a non-existent carcass. You can deter Foxes by sprinkling a harmless scent around the affected area. This is available in garden centres and DIY stores. You can also block holes in fences and place obstacles in the way of well-used paths.

•   **Foul play** Foxes, like dogs, scent mark their territory. A harmless artificial smell is an effective deterrent. Sprinkle it wherever there is a problem. Scrub away any Fox mess on your patio with a product that breaks down the residue (such as a biological washing power). This helps to prevent a repeat performance. You may have to persist in washing the same spot over several days, but the Fox will eventually get the message. Always use gloves when handling Fox waste.

**Top tips in brief**

•   **Do not feed Foxes** They will come back for more. Obviously.

•   **Protect your compost heap** Fence it in or use a compost bin.

•   **Make sure your refuse is secure** Dispose of edible waste in a secure bin and ensure that the lid is closed.

•   **Erect concrete building bases** Build your shed or garage on a concrete base to prevent Foxes from digging underneath it.

•   **Keep your garden clean and tidy** Overgrown gardens provide excellent shelter for Foxes, so clear these areas to remove the attraction. Tidy away shoes, gardening gloves and other items that might pique a Fox's curiosity.

A number of repellents, deterrents and Fox-proofing products are available. Be aware that it is an offence to use any product or chemical as a repellent unless it is labelled as such, and used in accordance with the instructions. Be aware also that many of these products are aimed at cats and may therefore upset your or your neighbours' pets. For more information visit www.foxolutions.co.uk.

**Above:** Keep your hutch doors bolted against hungry Foxes.

Interestingly, observations suggest that Foxes are most aggressive towards cats in rural areas, where there is more competition for food. In cities, night-time cameras show that many interactions between cats and Foxes remain stuck in neutral. The two animals go their separate ways, observing each other's space without much fuss. They sometimes even feed together from the same feeding station. Indeed, a Fox and a cat have been observed hunting voles side by side. Given that there are an estimated eight million domestic cats in the UK, outnumbering Foxes by around 40 to 1, it is likely that if there were any serious problem between the two animals it would surface a great deal more often.

**Above:** A Fox peers through the window with murder on its mind. Or perhaps just food.

# Homicidal Foxes?

Where the tabloids really have gone to town is when attacks involve human beings. Recent years have seen a small number of highly publicised attacks by Foxes on humans, all in south-east England. The circumstances of some have been hard to ascertain, with some suggestion of curiosity and panic on the part of the Fox rather than aggression. A handful, however, have been serious. One particular attack in June 2010, when nine-month-old twins Lola and Isabella Koupparis were both badly bitten in their cot by a Fox that entered the front door of their house in Hackney and climbed the stairs to the nursery, triggered a spate of headlines. It was widely suggested in the press that urban Foxes were growing bolder and more aggressive, that people – especially children – were at risk and that a cull was necessary to deal with the problem.

Foxes, like any wild animal armed with claws and teeth, are potentially dangerous, and any incident in which a person is harmed in any way is highly distressing for the victim and their family. Nonetheless, such incidents remain very rare, and scientists stress that it is important to get a sense of perspective. The behaviour of the Fox in the Koupparis incident – its readiness to enter the house and reluctance to flee when confronted by the

**Above:** Feeding a Fox by hand is never a good idea.

twins' mother – suggests that it may have been habituated in some way and had entered houses before. There are numerous incidents of Foxes that behave in a similar way, not with aggression, but simply with confidence – they may be found, for instance, curled up asleep on a sofa. Such incidents may involve Foxes that have been invited into neighbouring homes with food, habituated over time and even fed by hand. Allowing a Fox into a home and feeding it by hand is a bad idea under any circumstances.

As for the culling advocated by some sections of the media, while this might prove a brief panacea for public confidence all the evidence suggests that it would do nothing to reduce urban Fox numbers. New Foxes immediately inherit vacant territories, and studies have proven that more than 70 per cent of the Foxes in a given region would need to be culled before the population fell. An operation of this nature would be extremely expensive, logistically impractical and – once the shots started ringing out around suburbia – extremely unpopular. As plainly stated in a 2002 Bristol City Council leaflet entitled *Living with Urban Foxes*, 'Controlling urban foxes is difficult, expensive and never successful.'

## A bad press?

In recent years the UK media has played its part in inflaming public opinion against urban foxes. 'Parents tell of horror as fox attacks sleeping baby' (*London Evening Standard*, 10 Feb, 2013) was typical of a colourful reporting style, in which language such as 'living nightmare', 'menace', 'ferocious', 'savage' suggested that UK urbanites, their children and pets were living in significant danger from Foxes. 'Urban foxes are getting bigger... and more deadly,' warned another headline (*Daily Mail*, 3 Jan, 2011), illustrated with a photo of an especially large dead fox shot in Kent but providing no evidence to support either claim. Similar recent headlines have included 'Terror as Fox bites toddler, 3, on sledge, and tries to drag him away' (*The Sun*, 22 Jan, 2013) and 'Fox attack leaves 1 in 5 afraid of wildlife' (*Daily Telegraph*, 2 Sept, 2010). Mayor of London, Boris Johnson, also waded in: 'They may appear cuddly and romantic but foxes are also a pest and a menace, particularly in our cities' (*BBC News*, Feb 10, 2013). Reliable statistics, however, are hard to find in many such scare stories. A 2010 NHS survey of bites sustained across the UK cited a total of 8,500, of which 6,000 came from dogs, 1,875 from cats, 625 from humans and just 10 from Foxes. Thus a UK resident is on, average, 600 times more likely to be bitten by a dog, 187 times by a cat and 62 times by another human than by a Fox. Not quite such a good headline, perhaps.

# Live and let live

Another problem with trying to control urban Fox numbers – quite apart from the cost and logistics – is, simply, that many people like Foxes. Opinion polls conducted in both Germany and the UK have found that although some householders revile Foxes, more than half the respondents enjoy living with them. For some, the sight of a beautiful wild animal visiting their garden has a therapeutic effect and can provide emotional sustenance in times of need. The bloodcurdling midnight scream that has some people blocking their ears and calling 999 is, for others, a magical blast of wild nature in the heart of the city.

Love 'em or loathe 'em, urban Foxes are going nowhere. Our best option, then, is to learn to live with them. There are plenty of ways to deter them from damaging our gardens and of safeguarding our small pets. Meanwhile, perhaps we should count ourselves lucky that we have the privilege of watching this fascinating wild animal. Who needs expensive safaris to Africa when we have a wild predator on our doorstep?

**Above:** A fox at home in the back garden is a welcome sight for many householders.

**Below:** A memorable encounter with the wild, in the heart of the city.

# Foxes and People

Our feelings about Foxes run deep. Ever since we have co-existed with this animal it has captivated humankind in a variety of ways, whether as inspiration, resource, quarry, adversary or even pet. Chiefly, we have admired the Fox's ability to evade our clutches, imbuing it with human qualities – notably trickery, ingenuity and untrustworthiness – which are reflected in a wealth of literature and folklore. Indeed, so deeply is the Red Fox embedded in the rural culture of the British Isles that the Fox-hunting ban of 2004 threatened to divide a nation.

## The Fox in culture

The Red Fox has featured prominently in culture since ancient times. In the Old World it was widely depicted in the art of, among others, the ancient Egyptians and Greeks. In the Americas it has long been a totemic animal for many native peoples, one that has been attributed with magical properties: in North America the Apaches, for example, believed that the Fox passed fire as a gift to humankind. In ancient Japanese legend the multi-tailed Fox, or *Kitsune*, acted as a messenger from the spirit world capable of taking on human form; the more tails it had, the wiser and more powerful it was.

Early literary depictions of the Fox include the 'Fox and the Grapes', from *Aesop's Fables*, in which a Fox that finds itself unable to reach some grapes claims, disingenuously, not to have wanted them anyway – hence our expression 'sour grapes'. In the Bible, Jesus refers to Herod as a fox (Luke 13: 22), and Foxes appear in the Song of Solomon (2:15), which includes the well-known verse 'Catch for us the foxes, the little foxes that ruin the vineyards'.

This association of Foxes with trickery and untrustworthiness continues through literature, often in the form

**Opposite:** Mr Fox at *The Fantastic Mr Fox* press conference, before the 2009 film of Roald Dahl's celebrated book.

**Below:** Decoration from an Ancient Greek Attic red-figure vase, 5th century BC, depicting the Fox telling Aesop about animals.

**Above:** A Romanian stamp from 1996 featuring a Red Fox.

of the medieval character Reynard. Celebrated works featuring Foxes include Geoffrey Chaucer's *The Nun's Priest's Tale*, Machiavelli's *The Prince*, the fables of Jean de la Fontaine, Ben Jonson's *Volpone*, Joel Chandler Harris's *Uncle Remus* tales (featuring Br'er Fox) and Ted Hughes's poem *Thought Fox*, in which the Fox is a metaphor for an idea that sneaks up on the thinker. Children's literature is full of Foxes, from Beatrix Potter's *The Tale of Jemima Puddle-Duck* to Roald Dahl's *The Fantastic Mr Fox* and – more recently, Ali Sparkes's *Finding the Fox*, which explores the shamanic idea of shape-shifting that informed the mythology around this elusive animal in many early cultures.

Among countless other cultural depictions are Disney's *Robin Hood*, in which a Fox is, for once, the hero, and Leoš Janácek's opera *The Cunning Little Vixen*, inspired by a comic-strip story. Foxes also appear in everything from the emblem of Leicester City Football Club (nickname 'the Foxes') to the Audi Fox, of which some 1.1 million were sold worldwide in 1972–1979.

## The Fox in language

The impact of the Fox on culture is reflected in the way its name has permeated the English language. The word itself derives, via Old English, from the Proto-Germanic term *fuhsaz*. Today it has been adopted as a verb – 'to fox' – meaning to trick or fool by cunning or ingenuity, or in the passive form 'to be foxed', meaning to be baffled or confused. These meanings doubtless arise from frustrated farmers whose poultry runs or vineyards have been raided. In Australian slang, 'to fox' also means to pursue or tail stealthily. A further derived meaning is that of sexual attractiveness – usually applied to young women. Thus a 'fox' is the object of desire and 'foxy' a description of her qualities. 'Vixen', by contrast, is a less complimentary term, used to describe a quarrelsome or ill-tempered woman. As a balance to this sexism, the term 'silver fox' is also used to describe an alluring older man. More obscure meanings of 'fox' include the action of beer fermenting and of repairing a shoe by attaching a new upper. Fox is also the name of a Native American community who formerly lived along the Fox River, west of Lake Michigan, and describes the language of this people.

# Foxes for sale

The Fox's lustrous pelt, in both its red and silver forms, has been the envy of our species since we first started clothing ourselves. There is evidence that both Stone Age and Bronze Age peoples traded in Fox furs, and certainly the Romans wore Fox. Fur fell out of fashion in the UK during the 18th century, probably because of its associations with barbarism, but the trade resumed from the mid-19th century, fuelled by new incursions into the rich trapping territories of the north, notably Canada, Russia and the Pacific Northwest. Fox pelts were highly valued, along with the likes of Mink, Ermine and Beaver. During the early 20th century, more than 500,000 Fox skins were exported annually from Germany and Russia.

By the start of the 20th century, fur ranching was already providing a more efficient means of harvesting Foxes than trapping. The industry began in Alaska in 1890 with Arctic Foxes, and by 1930 Red Fox fur farms were established in the USA, Canada and Europe. Today fur farms account for some 85 per cent of the global fur

**Below:** Fox furs come in a variety of colours, of which silver is the most valuable.

**Above:** Model posing with fox fur.

**Below:** Employees capture Blue Foxes at a pelt farm near the village of Babino, Belarus. This farm produces some 2,400 Fox pelts annually.

trade. Fox fur is used for scarfs, muffs and jackets, and especially as trimmings for coats and evening wraps, with the pelts of Silver Foxes popular as capes. It takes approximately 14 Fox pelts to make a single full-length coat. North American Red Foxes are highly valued for their silky texture, especially those from Alaska.

During the latter half of the 20th century, the fur trade became the target of animal-welfare groups, and the anti-fur movement in the 1980s and '90s had a serious impact on the trade, which was declared illegal in the UK in 2000. Today fur farming continues over much of Europe, however, including in the Netherlands, Poland and Scandinavia, with Finland accounting for 81 per cent of European production. The International Fur Trade Federation argues that welfare issues have been addressed, and that Foxes today are treated and killed humanely. Wherever you stand on this contentious issue, there is no doubt that we have learned a great deal about Foxes from the many thousands kept on fur farms.

Fox has seldom featured on our menus. As a rule, the flesh of carnivores is too tough, too rich in protein and too plagued with parasites for human tastes. One exception is the Kazakh people of western Mongolia, who hunt Foxes with Golden Eagles (see page 85). There is nothing illegal about selling or eating Fox meat, however, and it occasionally appears in the window of UK speciality butchers, who import their farm-raised Foxes from Scandinavia.

# The Fox under attack

'War on the fox must continue to be the order of the day, and such warfare must be carried out by every means and with every weapon that is both practicable and humane.' These were the words of the UK Ministry of Agriculture in 1951, in a leaflet entitled *Wild Mammals and the Land*, and they have pretty much encapsulated the attitude of any farmer or landowner towards the Fox since time immemorial.

Records of culling Foxes, ostensibly to prevent them from taking livestock, date back to the 13th century. The practice has gained impetus over the years from such interventions as the Tudor culling acts of 1566 and the Fox Destruction Societies formed during the Second World War. In recent times attitudes have changed somewhat, with more legal controls and more humane techniques. Poisons were outlawed in England early in the 20th century, gin traps followed in 1954 (1973 in Scotland) and gassing with cyanide in 1986.

Today some 21,500–25,000 Foxes are culled annually in the UK (2000). This is considerably fewer than in some other countries, including Germany (600,000), Sweden (58,000), Finland (56,000) and Denmark (50,000). In Britain it takes a decision from a local council to have a particular Fox or Fox family condemned as vermin, and thus made eligible for destruction. The humane, legally acceptable way to do this is 'lamping'. This takes place at night and involves shining a beam in the Fox's eyes to dazzle it, then shooting it with a heavy-calibre weapon. A special lurcher-like breed of dog, called a longdog, is used to track down and retrieve any wounded animals.

**Above:** A pack of hounds tears apart their quarry.

**Below:** Red Fox dug out from its den by hunters in France.

# Tally ho!

Fox-hunting began as a form of pest control but over time developed into a sport and, for some rural communities, a way of life. The origins of the 'noble sport' lie in the Middle East. During medieval times Foxes were referred to, along with deer, as 'beasts of the chase'. The earliest known attempt to hunt a Fox with hounds was in Norfolk, in 1534. The first use of packs specifically trained to hunt Foxes was in the late 1600s, with the oldest Fox hunt thought to have been the Bilsdale in Yorkshire.

In its early days hunting with hounds was seen as the preserve of royalty. With the first Inclosures Act of 1773, which divided forests into smaller segments, it became harder to hunt deer and people increasingly turned their sights on Foxes. Soon Fox-hunting became so popular that large numbers of Foxes were imported to Britain – up to 1,000 per year during the mid-1800s. These came from the Continent, via Denmark, and sold for 15 shillings each to local hunts. Perversely, it has even been argued that the Fox, which was certainly in decline in Britain at this time, would have been eradicated were it not for Fox-hunting. Hunters, of course, always have an interest in preserving their quarry.

**Below:** For centuries, hunting with hounds was a mainstay of British rural culture.

'I've had people riding horses at me,' says Baz, recalling his time as a hunt saboteur – or 'sab' – during the 1990s. 'I've been chased with blokes with shovels, and one with a baseball bat.' Hunt sabbing during this period became a popular underground movement, through which people took direct action to express their opposition to Fox-hunting. Baz will not confirm where exactly his group operated but he explains how they would find out the details of a forthcoming hunt from *Horse and Hound* magazine and then meet to discuss strategy. The action took various forms, from throwing the hounds off the Fox's scent by using citronella and garlic spray, to physically sitting in the earth of a Fox gone to ground, denying access to the hounds or hunters.

Sabbing could be a dangerous business – not just because of the threats and violence meted out by angry landowners and members of the hunt (some of whom hired thugs to do their dirty work), but also the risk of falling the wrong side of the law, which could lose you your job. 'According to the press,' explains Baz, 'we were all drug-taking, cider-swilling, dreadlocked layabouts.' But sabs, he argues, came from all sections of society, united by the same principles and sense of injustice. It annoyed him that Fox-hunting was often depicted as universally popular among rural folk. 'I grew up on a farm, but my family have always been very anti-hunt,' he explains. 'These people were touting themselves as representing the countryside. I grew up there and I knew that they didn't.' Today, despite the Fox-hunting ban, sabbing continues, with social media allowing a whole new means of communication.

**Above:** Protestors gather at a Boxing Day Hunt meet in Fakenham, Norfolk.

In November 2004 the Hunting Act 2004 was passed after a free vote in the House of Commons – having already passed in Scotland in 2002. The law came into effect on 18 February 2005. It made the 'hunting of wild mammals with a dog' illegal in England and Wales, and sounded the death knell for traditional Fox-hunting, although certain modified forms of the sport are still legal. This new bill proved extremely controversial. The campaign drew strenuous arguments on both sides from such groups as the League against Cruel Sports (for the bill) and Countryside Alliance (against it), and pitted urban and rural communities against one another across the UK. The bill's supporters justified their case not in

**Above:** Police officers struggle to push back Countryside Alliance protesters during a demonstration outside The Houses of Parliament in London, 16 December 2002.

**Above:** Foxhounds are bred for their exceptionally keen sense of smell.

**Below:** Dead Red Foxes that were hunted in South Australia's Flinders Range as part of a state-sponsored programme to restore native fauna.

terms of conservation – the Fox is a common animal and Fox-hunting had never dented its numbers – but in terms of animal welfare. They argued that the stress and suffering endured by a hunted Fox was an unacceptable form of cruelty. Fox numbers have not risen significantly since the end of the ban.

Other countries in which people still hunt Foxes with hounds include the United States, Ireland, Canada, France, Italy, India and Australia, the sport often having been introduced from Britain. In 1730 rich American tobacco farmers introduced Foxes to their estates in Maryland and Virginia for the purpose of hunting. These Foxes have since been assimilated within the native Fox population (see page 31). In Australia, however, a similar scheme proved disastrous. The Red Fox is not indigenous Down Under, and ever since it was introduced to Melbourne by homesick British immigrants in 1855, also for the purpose of hunting, it has spread across two-thirds of the continent, driving numerous native species into decline and even extinction (see page 33), and causing serious losses to sheep farmers. In the state of Victoria today, Fox-hunting with hounds accounts for some 650 Foxes annually – compared with over 90,000 shot by those after a government bounty.

# Foxes as friends

For all the people who want to hunt, shoot or gas Foxes, there are at least as many who love them and want to get as close to them as possible – even to the extent of keeping them as pets. There is some evidence that Foxes may have been kept as domestic animals in ancient Egypt, and even buried with their masters.

Today some people take Fox cubs from the wild and raise them as household pets. While such people may have the best of intentions – perhaps trying to provide a home for an abandoned cub – and the young Foxes may be delightful company for a while, such stories seldom have happy endings. Foxes are, ultimately, wild animals. They are destructive and smelly and, once these traits have stretched householders beyond endurance, the unfortunate pets are often abandoned to their fate. Wild Foxes are definitely better off left where they belong: in the wild.

**Above:** Foxes can have domesticity bred into them.

That said, there have been some fascinating experiments conducted into the domestication of Foxes, notably in Russia during the 1950s using Silver Foxes from fur farms. They were selectively bred, the researchers selecting individuals from a litter that were friendliest towards humans, breeding from those individuals, then breeding from their friendliest cubs, and so on. By the sixth generation the animals were behaving like domestic dogs, craving human company, bonding with people and responding to human cues and moods. To the researchers' amazement they also started to develop new physical traits, including shorter muzzles, curlier tails and floppier ears – just as happened with domestic dogs on their journey from their Wolf ancestors. Such Foxes are sold as family pets, mostly in the US, where they fetch upwards of £4,000.

**Below:** An orphaned red fox cub shows no difficulty feeding alongside her feline companions.

# Watching Foxes

Catching sight of a Fox is always thrilling. After all, we do not – in the UK at least – have a surfeit of genuine wild predators roaming our land. For those who live in towns, where Foxes are bolder and more habituated to people than they are in the countryside, this thrill comes along rather more often than for those who live in rural areas. A Fox sighting is generally brief and unexpected. The animal may suddenly trot around a corner or across a road, only to disappear by the time you have clocked what you are looking at. Your glimpse may be no more than a shadow in a car park, or a streak of russet at the back of a field. With time, experience and patience, however, you can give yourself a better chance of finding Foxes and watching them at your leisure.

# In the countryside

It may seem counter-intuitive, but there is no denying that today's average rural dweller sees Foxes much less often than the average town inhabitant. The UK countryside is still home to some 85 per cent of the nation's Foxes, but rural individuals have much larger home ranges than their urban cousins, and they behave with all the wariness of wild animals that depend for their survival upon the stealth needed to capture a Rabbit or dodge an angry gamekeeper.

Your best chance of watching Foxes for any decent length of time is at the earth during the breeding season. This is around late April/early May, when the kits are venturing above ground for the first time and the adults are out and about capturing food. If you find an active Fox earth, you can enjoy wonderful sightings of the kits at play and interacting with the adults.

The trick when watching a Fox's earth is to find a concealed viewing spot that is not too close to the earth and is downwind from the Foxes – in other words, the

**Opposite:** A fine view of a wild Red Fox going about its business is always a thrill.

**Below:** Observing cubs outside the earth is the holy grail of any Fox watcher.

breeze should be blowing in your face, rather than from you towards the Foxes. Try to get to the viewing location before dawn and stay as long as you can – all day, if possible. The cubs will explore further afield when the parents leave them alone. This is a good time to take photographs, but keep an eye out for the parents, which are more wary and will be coming and going with food. There may also be a helper or two hanging around with the youngsters. After watching for a while you will get to know the adults' favoured approach routes. Do not press your luck: a vixen may move her cubs to a different earth if she knows that people are around. Also keep the location to yourself; not everyone likes Foxes as much as you do.

A Rabbit warren can also be a good place to watch Foxes. When you find an active warren, look out for scats, tracks and other signs in the immediate vicinity. If there are Foxes about (and there usually are), they will probably visit the warren regularly in the hope of snatching an unwary victim. Again, early spring is the best season, when the adults are providing for more hungry mouths than during the rest of the year, and dawn or dusk are the best times to watch. Get yourself in position early – somewhere comfortable, not too close and downwind of the warren – and watch. Serious Fox watchers and wildlife photographers sometimes construct tree hides. Not only are they concealed in this high vantage point, but also their scent tends to drift over a

**Above:** Keep a regular watch on Rabbit warrens if you want to see a hunting Fox.

**Below:** For the serious rural Fox watcher or photographer, an elevated hide makes an excellent vantage point.

Fox's head and there is a better view of the insides of hollows or over hedges.

Other productive habitats for Fox watching include cattle pasture, where the disturbed ground and its smattering of cowpats provides ideal conditions for earthworms, and areas of long, rank grass, where voles and other rodents have their runways. If you chance upon a Fox in such a place, it is usually worth returning: Foxes visit favourite hunting grounds within their home range on a regular rotation. Do not give up if the weather turns cold; a fresh snowfall can make it easier to find and see Foxes, which leave quite obvious tracks and may come out into the open as they sniff about and pounce on prey hidden beneath the blanket of white.

Ask around to find out about good Fox-watching places near you, or visit a local nature reserve – such as one managed by the RSPB or Wildlife Trusts – and get advice from the warden or volunteers. They may be able to direct you somewhere, perhaps a hide, where Foxes are often seen. Even if you do not find a Fox, you are bound to come across other wildlife. A quiet vigil over a field or Rabbit warren at dusk may produce a Badger or Barn Owl, while an early spring morning spent watching a Fox earth is a wonderful way to enjoy the dawn chorus. Sitting still in one place for a while allows all kinds of wildlife to accept your presence and adapt to it, and is an opportunity to watch normally elusive species, from Wood Mice to Common Whitethroats.

**Above:** Sitting quiet and still at dusk may reward you with sightings of other animals, such as this Barn Owl.

**Below:** Cattle pasture is good Fox habitat; the trampled ground and cowpats harbour abundant earthworms.

# Tracks and signs

You do not need to see a Fox to know that it is there. Like all animals, Foxes leave evidence of their comings and goings all around their territory. Your first clue may be smell: the distinctive musky odour that tells you a Fox has passed this way and sprayed its scent nearby. Once this assaults your nostrils, it is worth looking for the animal's other calling cards.

**Above:** Fox tracks are easily spotted in snow.

## Footprints

A Fox's tracks usually occur in a long straight line, like a running stitch, and follow a regular trail used again and again – often along territorial boundaries, such as beside a hedge, and leading to key access points, like a gap in the hedge. The hind feet typically tread directly in the tracks of the forefeet (a habit known as direct registering), so that the animal may appear to have only two feet. The paw prints themselves measure about 4–6cm (1½–2½in) long by 3.5–5cm (1½–2in) wide, and show four roughly round toes in front of a small, roughly triangular pad, with the mark of a claw in front of each toe visible on most surfaces.

Fox prints are easily distinguished from those of a Badger (which has five toes and broader, longer claws) and a cat (no claws). They are trickier to tell from a domestic dog's, which are similar in form though generally a bit broader. However, on a good surface, a Fox track should show an impression of hair on the underside and a distinct chevron mark across the pad, both of which are absent from a domestic dog's track. Dog tracks vary greatly in size according to the breed, and are often far larger than a Fox's. Early morning after a fresh snowfall is a good time to look for Fox tracks, but bear in mind that tracks may appear larger in snow.

**Above:** A Fox track shows four round, clawed toes in front of a triangular pad.

## Scats

Look out, too, for a Fox's droppings – known as scats. You will often find these on a prominent raised feature, such as tussock of grass or a molehill, where the Fox has deposited them as a conspicuous territorial marker.

**Above:** Badger tracks show much longer claws than those of a Fox or dog.

A newly mowed meadow is another favoured location. Fox scats are usually about 2cm (¾in) across by 3–9cm (1–3½in) long, ropey, slightly twisted and tapering at one end. They have a musky odour – not as offensive as those of a dog. Look closer and you may see the remains of teeth, fur and bones. In summer the droppings are a pale whitish-grey owing to the calcium from the bones of all those small mammals eaten at that time of year. At other times they may be darker, and in late summer they are often blackish and more jelly-like in texture owing to all the blackberries the Fox has consumed.

**Above:** Fox droppings are darker when full of blackberries.

## Other signs

The experienced tracker may come across other signs of Fox activity. A place where Foxes lie up is known as a lay and may be visible as a small hollow in the grass or snow, with a few hairs pressed into the indentation and tracks leading in and out. A bird killed by a Fox and consumed on the spot will have the primary wing feathers sheared through near the base, where the Fox bit them out in clumps. This contrasts with the remains of a bird killed by a raptor such as a Sparrowhawk, from which the primary feathers are plucked out individually, sometimes showing a small scar near the base of each quill left by the raptor's hooked bill.

**Above:** The head of a domestic chicken buried in soil is gruesome evidence of a fox kill.

**Below:** A Red Fox with a dead Red-legged Partridge.

# In town

Finding Foxes in town is a great deal easier than in the country. Tracking them may be tricky – Foxes do not leave footprints on hard pavements, and you cannot climb over your neighbours' fences to search their gardens for scats – but the animals occur in much greater densities, have smaller territories and are so unconcerned about humans that they show themselves much more readily here than in the countryside.

The easiest place to watch a Fox for a decent length of time – as opposed to simply glimpsing one trotting past – is in the sanctuary of your own garden, if you have one, or perhaps the grounds of your block of flats. If you are lucky enough to have Foxes move in beneath your shed, you may be able to watch the antics of the cubs and labours of the parents from a kitchen or bathroom window. The Foxes will be wary of noise and disturbance, however, so are likely to be more active when fewer people are around.

**Below:** There is no easier place to watch a Fox than in your own back garden.

Other good places to look in built-up areas include cemeteries, where Foxes find a quiet retreat and may dig their earth beneath gravestones, and landfill sites, where they may make regular foraging visits. Also keep an eye out from the train during your morning or evening commute: Foxes make their earths in railway embankments, and often appear beside the tracks or even on station platforms.

**Above:** Keep an eye out for Foxes along railway lines.

**Below:** An open skip on a city street is bound to attract the attention of a Fox or two.

# To the rescue

For some Foxes that fall victim to misfortune, whether through accident, injury or – in the case of a cubs – losing parents, help is at hand in the form of rescue centres. The UK's leading animal welfare charity, whose work includes the rescue, rehabilitation and release of wild animals – including Foxes, is the Royal Society for the Prevention of Cruelty to Animals (RSPCA). In 2013 its four wildlife centres around the UK took in between them 15,254 sick, injured or orphaned wild animals, ranging from Fox cubs to birds of prey. These centres are equipped with veterinary surgeries, orphan wards and various paddocks, pools, pens and aviaries to accommodate the many species they admit. The RSPCA stresses that Wildlife rehabilitation is extremely tricky and best left to experts. If you do come across an injured or orphaned Fox, you can contact them on 0300 1234 999. Find out more at www.rspca.org.

In addition to the RSPCA, numerous smaller organisations and rescue centres also care for Foxes and other wildlife around the UK. The National Fox Welfare Society (www.nfws.org.uk) is a voluntary organisation dedicated to helping the Red Fox. Its services include providing rescues for sick and injured Foxes and free treatments for Foxes suffering from mange. More regional charities

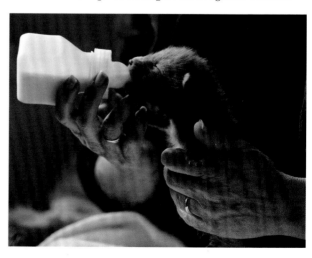

**Right:** One-week-old Fox cub being bottle-fed at the Folley Wildlife Rescue in Kent, England.

range from The Fox Project (www.foxproject.org.uk), which provides a wildlife ambulance service in the Southeast and always aims to return adult Fox casualties to their home territory, to Secret World Wildlife Rescue (wwww.secretworld.org), which runs a 24/7 year-round wildlife rescue service around the West Country from its base in Somerset, and The Sanctuary Wildlife Care Centre (www.wildlife-sanctuary.co.uk), which runs a similar service in Northeast England from its base in Morpeth, Northumberland. All these organisations also offer extensive advice on living with foxes and humane methods of Fox deterrence. Many also have opportunities for volunteers.

**Above:** Fox cub at The Fox Project rescue centre, Kent, prior to re-introduction to the wild.

**Below:** Orphaned Red Fox cubs at a rehabilitation centre in London.

# Glossary

**Bergmann's rule** Principle first described by 19th-century German scientist Carl Bergmann, which states that animals within a given genus – or different populations within a single species of animal – are larger in colder environments than in warmer ones. His theory was that larger animals have a lower surface area to volume ratio than smaller animals, so they radiate less body heat per unit of mass and therefore stay warmer in cold climates.

**Boreal forest** Subarctic area of northern Eurasia and North America located just south of the tundra and covered largely with coniferous forests dominated by firs and spruces. Also known as the taiga, this biome occupies about 17% of the earth's surface.

**Canidae** The dog family: a lineage of carnivorans that comprises around 37 species (including wolfs, foxes and jackals), and is one of around 23 families that make up the order Carnivora.

**Carnassial teeth** Large teeth found in most carnivorous mammals, including foxes, that are used for shearing meeth in a scissor-like way. They are set far back in the mouth, and are the modified fourth upper premolar and the fourth lower molar.

**Convergent evolution** The independent evolution of shared or similar features in species of different evolutionary lineages. Thus birds and bats have evolved flight independently but use them for the same purpose. The opposite of this phenomenon is divergent evolution.

**Copulatory tie** Position, also known as 'coital tie', in which foxes and other members of the dog family become physically locked together during mating. It is caused by the swelling of the male's reproductive organs inside the female's. Such a tie may last from five minutes to one hour.

**Gekkering** High-pitched, rapid, stuttering call given by Red Foxes during disputes or encounters with territorial rivals.

**Guard hairs** Long coarse hairs in the coats of many mammal species, including Foxes, which form a separate layer that protects the soft underfur.

**Holarctic** Biogeographic region that comprises the northern areas of the northern hemisphere. It is divided into the Palearctic, in Europe and Asia, and the Nearctic, in North America.

**Implantation** Part of the reproduction process in placental mammals, in which the fertilised egg becomes attached ('implants') to the lining of the uterus.

**Lamping** Method by which Foxes and other nocturnal mammals are hunted. A powerful spotlight ('lamp') is shone in the animal's eyes, enabling the hunter to locate the target by its eyeshine and shoot it. A Fox will generally continue to stare at the light and thus allow a close approach, often by an offroad vehicle. Lamping is regarded as the most humane method of culling foxes.

**Mate guarding** The period of time, while a female fox is in oestrus, during which her mate accompanies her constantly in order to prevent her mating with other males.

**Morph** A distinct form within a species that differs from other forms within the same species occupying the same habitat at the same time. The Silver Fox, for example, is a colour morph of the Red Fox.

**PCBs** Short for polychlorinated biphenyls, these are synthetic organic chemicals that were formerly widely used as coolant fluids and in other industrial and domestic applications. The toxic effect of PCBs in the environment has caused their production to be banned in many countries.

**Sarcoptic (canine) mange** A serious skin condition, known simply as mange, that afflicts Foxes and other animals and is caused by the contagious infestation of the burrowing mite *Sarcoptes scabiei canis*. Mange causes intense itching and crusting of the skin, leading to hair loss and ultimately, where untreated, death.

**Sexual dimorphism** Physical differences between the male and female of a species, typically of size and/or colouration.

# Further Reading and Resources

## Books: factual

Harris, Stephen, *Urban Foxes* (Whittet Books, 2000)
Fascinating study that dispels many myths about urban Foxes; second edition updated with new information on social interactions.

Hemmington, Martin, *Fox Watching: In the Shadow of the Fox* (Whittet Books, 2014)
Handbook on how to watch Foxes in the wild.

Macdonald, David, *Running with the Fox* (Harper Collins, 1989)
Classic, comprehensive and highly readable study of the Red Fox by one of the world's foremost authorities. Includes an interesting chapter on the habits of Red Foxes in Israel, comparing them to their British cousins. A bible for Fox-lovers.

Wallen, Martin, *Fox* (Reaktion Books, 2006)
Examines the role of Foxes among different cultures around the world.

## Books: Children's fiction

Dahl, Roald, *Fantastic Mr Fox* (Puffin; Re-issue edition, 2014)
Colour edition of one of Roald Dahl best-loved stories. Clever, handsome Mr Fox thwarts the farmers to save his family.

Green, Alison, *The Fox in the Dark* (Scholastic UK, new edition 2014)
Classic picture storybook, with poetry, humour and lively illustrations.

Morpurgo, Michael, *Little Foxes* (Egmont, 2013)
Gripping animal story from one of Britain's best-loved children's authors, which tells the story of a bullied child who befriends a wild fox family.

Sweeney, Matthew, *Fox* (Bloomsbury 2002)
Children's novel about a boy, struggling at a new school, who strikes up a friendship with a homeless man and his pet fox.

## Online

BBC www.bbc.co.uk/nature/life/Red_Fox
Information, images and video, including clips from *Springwatch* and other BBC series

National Geographic http://animals.national geographic.com/animals/mammals/red-fox
Extensive resource from *National Geographic*, with information, images and video

Wildife online: Natural History of the Red Fox www.wildlifeonline.me.uk/red_fox.html
Excellent resource, with extensive, detailed and up-to-date information and advice on all aspects of the Red Fox. Compiled by UK naturalist Mark Baldwin, with an extensive team of expert editors and contributors.

## Conservation and care

The Mammal Society (www.mammal.org.uk)
Campaigning charity that supports conservation and research for mammals across Britain and Ireland. Offers extensive information and advice about the Red Fox.

National Fox Welfare Society (www.nfws.org.uk)
Voluntary organisation dedicated to helping the Red Fox. Rescues sick and injured Foxes and provides free treatments for Foxes suffering from mange.

RSPCA (www.rspca.org.uk)
The UK's largest animal welfare charity. Four regional centres run wildlife rescue and rehabilitation programmes for sick, injured or orphaned wildlife, including Red Foxes.

RSPB (www.rspb.org.uk/)
The RSPB is the country's largest nature conservation charity. Manage reserves around the UK to save birds, mammals, including Red Foxes, and habitats.

Wildlife Rescue Centres Nationwide (www.helpwildlife.co.uk/southeast.php)
Inventory of all charities and organisations in the UK offering rescue and rehabilitation services for wild animals, including Red Foxes.

Wildlife Trusts (www.wildlifetrusts.org)
National conservation charity comprising 47 separate regional wildlife trusts across the UK, with information and advice on all aspects of British wildlife, and an extensive network of reserves on which you can watch wild Foxes.

# Image credits

**Key** t=top; l=left; r=right; tl=top left; tcl=top centre left; tc=top centre; tcr=top centre right; tr=top right; cl=centre left; c=centre; cr=centre right; b=bottom; bl=bottom left; bcl=bottom centre left; bc=bottom centre; bcr=bottom centre right; br=bottom right

FLPA: Frank Lane Photography Agency; NPL: Nature Picture Library; G: Getty; SH: Shutterstock; RSPB: RSPB-Images.com

**Front cover and spine** SH; **back cover** t David Pressland/FLPA, b Jules Cox/FLPA; **half title** SH; **title** SH; **4** SH; **5** ImageBroker/Imagebroker/FLPA; **6** tl Dave Pressland/FLPA, map: Julie Dando/Fluke Art; **7** SH, br Yossi Eshbol/FLPA; **8** t, b SH; **9** Igor Shpilenok/NPL; **10** Fabrice Cahez/Biosphoto/FLPA, SH; **11** Igor Shpilenok/NPL, SH; **12** tr Michael Quinton/Minden Pictures/FLPA, bl Inge van der Wulp/Minden Pictures/FLPA, br Dave Pressland/FLPA; **13** c Hanne & Jens Eriksen/NPL, b SH; **14** Desmond Dugan/FLPA; **15** SH; **16** t, b SH; **17** t Michael Quinton/Minden Pictures/FLPA, SH; **18** cl Jules Cox/FLPA; b John Hawkins/FLPA; **19** t Fabrice Cahez/NPL, c, b SH; **20** SH; **21** Mike Unwin;

**22** t Natural History Museum (WAC)/NPL, c, b SH; **23** l SH, r Steve Gettle/Minden Pictures/FLPA; **24** map: Julie Dando/Fluke Art; **25** Mike Unwin, bl, br SH; **26** Andrew Parkinson/RSPB; **27** t SH, ct Jenny Hibbert/RSPB, cb SH, b Kevin Schafer/Minden Pictures/FLPA; **28** t Jim Brandenburg/Minden Pictures/FLPA, ct, cb, SH, b Dean Bricknell/RSPB; **29** t SH, c Sergey Gorshkov/NPL, b Steven Kazlowski/NPL; **30** SH; **31** Bill Draker/Imagebroker/FLPA; **32** t Michael Quinton/Minden Pictures/FLPA, b TBC; **33** bl Roland Seitre/NPL br Mike Unwin; **34** Gerard Lacz/FLPA; **35** SH; **36** tl SH, Laurent Geslin/NPL; **37** t Paul Hobson/FLPA, b Fabrice Cahez/Biosphoto/FLPA; **38** t Fabrice Cahez/Biosphoto/FLPA, c SH, b Frederic Desmette/Biosphoto/FLPA; **39** t, b SH; **40** t Martin H Smith/FLPA, c Fabrice Cahez/NPL, b SH; **41** t Bruno D'Amicis/NPL, b Laurent Geslin/NPL; **42** t Dieter Hopf/Imagebroker/FLPA, b SH; **43** t, b SH; **44** t, b SH; **45** t Laurent Geslin/NPL, b TBC; **46** Fabrice Cahez/Biosphoto/FLPA; **47** b Dieter Hopf/Imagebroker/FLPA; **48** Charlie Summers/NPL; **49** Laurent Geslin/NPL; **50** tl David T. Grewcock/FLPA, tr SH, bl Martin H Smith/FLPA, br SH; **51** Sergey Gorshkov/NPL; **52** A©Biosphoto, Fabrice Cahez/Biosphoto/FLPA; **53** t Laurent Geslin/NPL, b Laurent Geslin/NPL; **54** Laurent Geslin/NPL, **55** tr SH, bl Laurent Geslin/NPL; **56** Andrew Cooper/NPL; **57** cl Klaus Echle/NPL, bl Sam Hobson/NPL, br Kevin J Keatley/NPL; **58** Jules Cox/FLPA; **59** b SH; **60** t Laurent Geslin/NPL; b T.J. RICH/NPL;

61 t SH, b Elliott Neep/FLPA; 62 t, b SH; 63 t SH, b Karim Smaoui/Biosphoto/FLPA; 64 Imagebroker/FLPA; 65 Fabrice Cahez/NPL; 66 t Laurent Geslin/NPL, b Laurent Geslin/NPL 67 t Laurent Geslin/NPL, br John Veltom/FLPA; 68 t Laurent Geslin/NPL, b Gary K. Smith/NPL; 69 t Fabrice Cahez/NPL, b T.J. RICH/NPL; 70 t Colin Seddon/NPL, b Laurent Geslin/NPL; 71 Shattil & Rozinski/NPL; 72 Konrad Wothe/Minden Pictures/FLPA, 73 t Dick Hoogenboom/Minden Pictures/FLPA, b Laurent Geslin/NPL; 74 David Pattyn/NPL; 75 cl Laurent Geslin/NPL, cr Fabrice Cahez/Biosphoto/FLPA, bl Loic Poidevin/NPL, br Laurent Geslin/NPL; 76 ImageBroker/Imagebroker/FLPA; 77 t Imagebroker/FLPA, b Andrew Parkinson/NPL; 78 SH; 79 Martin B Withers/FLPA; 80 t SH, b Hannu Hautala/FLPA; 81 t Laurent Geslin/NPL, b Wayne Hutchinson/FLPA; 82 t David T. Grewcock/FLPA, b SH; 83 t Paul Sawer/FLPA, c Sergey Gorshkov/Minden Pictures/FLPA, b SH; 84 Richard Costin/FLPA; 85 c Vincent Munier/NPL, b Pete Oxford/NPL; 86 t Armin Floreth/Imagebroker/FLPA, b Elliott Neep/FLPA; 87 t SH, b Laurent Geslin/NPL; 88 t Colin Seddon/NPL, b SH; 89 bl SH, br Laurent Geslin/NPL; 90 Uri Golman/NPL; 91 t John Hawkins/FLPA, b Will Burrard-Lucas/NPL; 92 Jamie Hall/FLPA; 93 Duncan Usher/Minden Pictures/FLPA; 94 tl Mike Read/RSPB, tc Laurent Geslin/NPL, tr Laurent Geslin/NPL, b Ray Bird/FLPA; 95 Mike Unwin; 96 t Bruno D'Amicis/NPL, b Florian Möllers/NPL; 97 t Roger Tidman/FLPA, b Shattil & Rozinski/NPL; 98 t Andy Rouse/NPL, b Paul Hobson/FLPA; 99 Laurent Geslin/NPL; 100 ARCO/NPL; 101 Warwick Sloss/NPL; 102 Terry Whittaker/FLPA; 103 t Terry Whittaker/FLPA, b Laurent Geslin/NPL; 104 Vera Anderson/G; 105 De Agostini Picture Library/G; 106 SH; 107 SH; 108 t SH, b Viktor Drachev/AFP/G; 109 t Bryan and Cherry Alexander/NPL, b Gerard Lacz/FLPA; 110 b SH; 111 t Martin H Smith/NPL 01077479, b Adrian Dennis/Staff/G; 112 t, b SH; 113 tr Angelo Gandolfi/NPL, b Angelo Gandolfi/NPL; 114 Dale Sutton/RSPB; 115 Laurent Geslin/NPL; 116 c Dieter Hopf/Imagebroker/FLPA, b Imagebroker, Raimund Kutter/Imagebroker/FLPA; 117 t Dave Pressland/FLPA, b SH; 118 t David Tipling/FLPA, c ImageBroker/Imagebroker/FLPA, b Michael Durham/FLPA; 119 t John Veltom/FLPA, c R P Lawrence/FLPA, b Gerard Lacz/FLPA; 120 SH; 121 t SH, b Laurent Geslin/NPL; 122 Jules Cox/FLPA; 123 t Mark Sisson/RSPB, b Laurent Geslin/NPL; 126 Igor Shpilenok/NPL

ACKNOWLEDGEMENTS

# Acknowledgements

I would like to thank all those whose help and encouragement helped bring this book to fruition. It has been a pleasure, as ever, working with the Natural History team at Bloomsbury. I am especially grateful to to Julie Bailey, for presiding over the excellent Spotlight series and developing this title; to Jasmine Parker, for all her editorial energies and resourcefulness; and to Nigel Redman, for his valuable input and support. Thanks, too, to Susan McIntyre, for her excellent layout and design.

I would also like to thank all those naturalists and conservationists whose dedicated study of the Red Fox over the years has brought us a deeper understanding of this remarkable animal and helped us learn to live with it. Among the many writers and film-makers who have inspired and informed me, I am especially grateful to Mark Baldwin, whose excellent website Natural History of the Red Fox (www.wildlifeonline.me.uk/red_fox.html) proved an invaluable resource during the writing of this book.

Finally, I would like to thank my family: my parents, who encouraged my love of nature from the earliest days; and my wife Kathy and daughter Florence, who have put up with my long book-writing absences and with whom I have shared many of my most memorable moments with foxes and other wildlife.

# Index